A Geometric Analysis of the Platonic Solids and Other Semi-regular Polyhedra

With an introduction to the phi ratio

For Teachers, Researchers, and the Generally Curious

by

Kenneth J. M. MacLean

A Geometric Analysis of the Platonic Solids and Other Semi-regular Polyhedra

Cover art was provided by George W. Hart. Visit the Encyclopedia of Polyhedra at http://www.georgehart.com/virtual-polyhedra/vp.html

Another Big Picture book

    Library of Congress Cataloging-in-Publication Data

MacLean, Kenneth James Michael, 1951 –
A Geometric analysis of the platonic solids and other semi-regular polyhedra : with
an introduction to the phi ratio : for teachers, researchers, and the generally
curious / by Kenneth James Michael MacLean
p. cm. – (Geometric exploration series ; 1)
Includes bibliographical references and index.
ISBN-13: 978-1-932690-99-6 (hardcover : alk. paper)
1. Polyhedra–Models. 2. Geometry–Study and teaching. 3. Golden section. I. Title.
QA491.M33 2007
516'.156–dc22
2007000763

2nd Printing – February 2015

The Big Picture is an imprint of Loving Healing Press

Acknowledgments

This book is dedicated to those who can appreciate the logic of numbers and the beauty of nature, for they are both aspects of the same unifying principle.

A special acknowledgment goes to Rick Parris for his excellent free software, WinGeom, with which the drawings in this book were made. You can find Rick's programs at http://math.exeter.edu/rparris/

Contents

Introduction

In this book I present a detailed analysis of the five Regular Solids, along with five other important polyhedra:

- The Cube Octahedron
- The Rhombic Dodecahedron
- The Rhombic Triacontahedron
- The Icosa Dodecahedron
- The Star Tetrahedron

We will also thoroughly investigate the pentagon, which is nature's key to unlocking the secrets of the Platonic Solids and other important polyhedra. The pentagon is composed entirely of Phi relationships. It turns out that knowledge of Phi is indispensable in the understanding of these polyhedra. This book explains the geometric basis of the Phi ratio and the Fibonacci series, and how they are derived mathematically. For those readers who are serious about sacred geometry, these concepts are essential.

For each polyhedron we calculate:

- Volume
- Surface area
- Central angles to all vertices
- Surface angles
- Dihedral angle
- Centroid to vertex
- Centroid to mid–edge
- Centroid to mid–face
- Side to radius
- Internal planes and their relationships

Volume and surface area is calculated in terms of the edge length of each polyhedron, and also, for comparison purposes, within the unit sphere of radius 1. This sphere is the sphere that contains the outermost vertices of the polyhedron. For the 5 regular solids, all vertices lie on the unit sphere.

These two calculations help us to visualize the polyhedra in relation to each other.

Each polyhedron has an accompanying reference chart.

Appendix A combines the data for all of the polyhedra into one large chart.

A note on dihedral angles:

The dihedral angle is the intersection of two planes, like so:

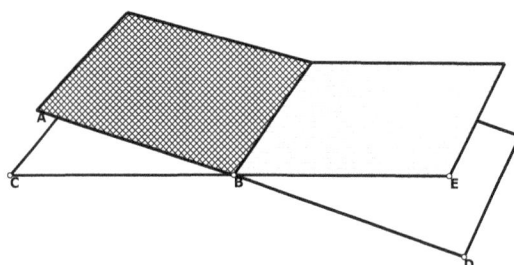

The dihedral angle is usually considered to be the angle ABC. In this book, however, we will use the angle ABE (or CBD), because it is easier to see when building 3 dimensional models. I encourage all readers to build these polyhedra with the Zometool, for it will lead to much greater comprehension of the material.

In this book we avoid trigonometry unless calculating central and dihedral angles. Geometry comes from the words "geo," or earth, and "metrien," to measure. Geometry is about the real and the tangible. It is tempting, when one knows an angle and a side length, to simply use trig to get the other side. One may use the Law of Cosines and other tricks to quickly get distances. Such reasoning does not dig deeply enough, however, and often obscures the geometric detail. In my mind, geometry is all about discovering relationships. In this book I have built models of every solid I have analyzed, allowing me to delve very deeply into the ratio and proportion of these polyhedra. I have found that if one studies long enough, one can always come up with quantities using only right triangles and the Pythagorean Theorem. Such an approach may seem long-winded and unnecessary to some, but it generates a tremendous amount of useful data, and a ton of learning. As researchers we can appreciate the data, and as teachers we are always looking for ways to keep mathematics interesting and, dare I say, even exciting for our students.

Make no mistake, there is a lot of exciting stuff in this book for those who can appreciate the beauty of numbers. These polyhedra are merely reflections of Nature herself, and a study of them provides insight into the way the world is structured. Nature is not only beautiful, but highly intelligent. As we explore the polyhedra in this book, this will become apparent over and over again. So crack open the book and get started on a stimulating adventure into the world of geometry!

Part I

We begin in Part One with the Tetrahedron, the Octahedron, and the Cube. These polyhedra are all based on the $\sqrt{2}$ and the $\sqrt{3}$, unlike their much more complex brethren, the Icosahedron and the Dodecahedron, which are based on the $\sqrt{5}$ and the Phi ratio.

Chapter 1

The Tetrahedron

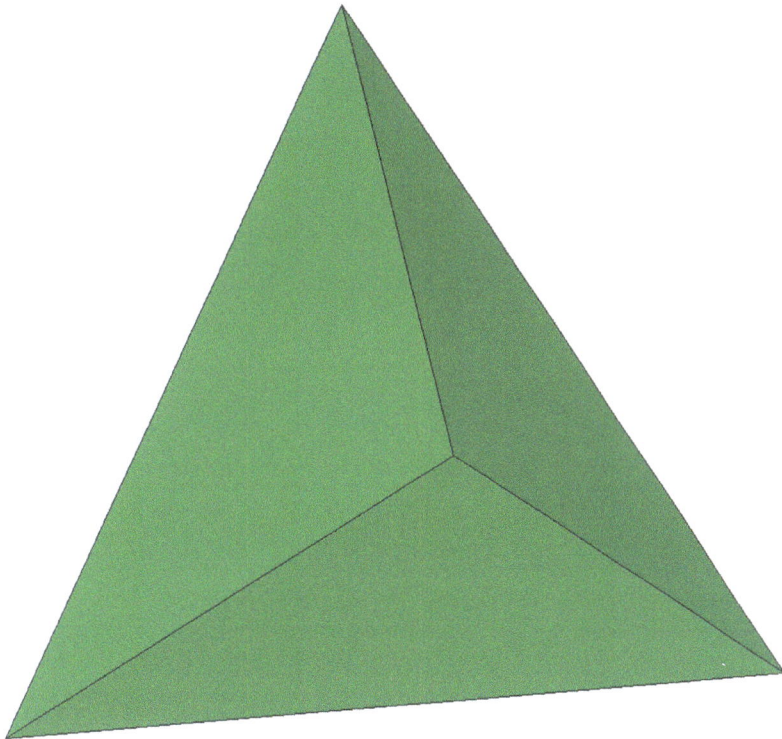

The tetrahedron has 6 sides, 4 faces, and 4 vertices. In Figure 1.1 below the base is marked out in gray: the triangle BCD. Each of the faces is an equilateral triangle.

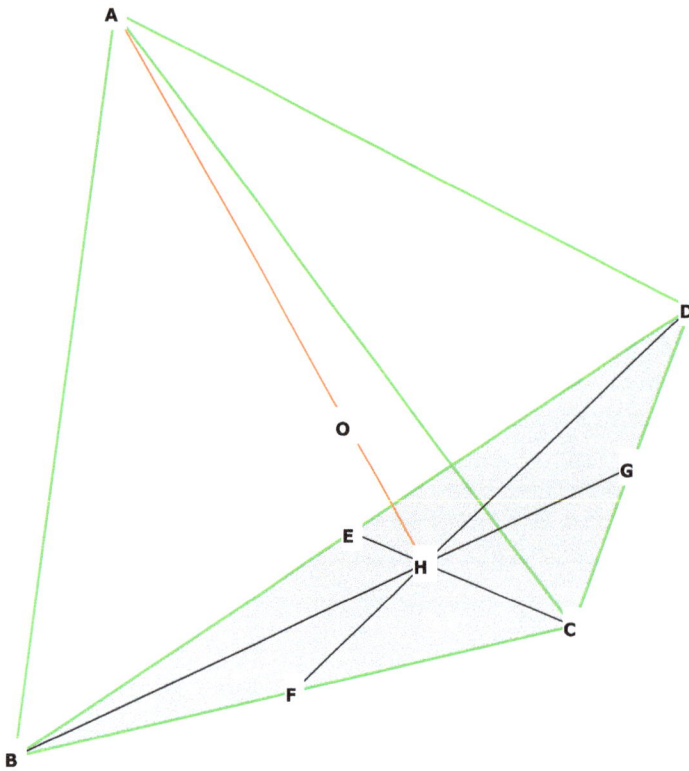

Figure 1.1: Tetrahedron, interior

Each side of the tetrahedron is in green. We will refer to the side or edge of the tetrahedron as ts.

1.1. Height of Tetrahedron

From *The Equilateral Triangle* we know that:

Area $\triangle BCD = \frac{\sqrt{3}}{4}\, ts^2$.

Now we need to get the height of the tetrahedron, $\overline{AH}$.

From *The Equilateral Triangle* we know that:

$\overline{BH} = \frac{1}{\sqrt{3}}\, ts$.

Now that we have $\overline{BH}$, we can find $\overline{AH}$, the height of the tetrahedron. We will call that h.

$$h^2 = \overline{AB}^2 - \overline{HB}^2 = 1ts - \left(\frac{1}{\sqrt{3}}ts\right)^2 = \frac{2}{3}\, ts^2.$$

$$h = \frac{\sqrt{2}}{\sqrt{3}}\, ts = 0.816496581\, ts. \tag{1.1}$$

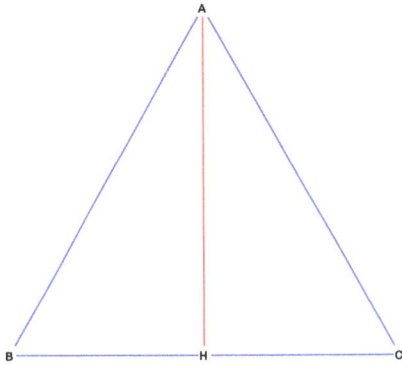

Figure 1.2: Height of equilateral triangle

1.2. Volume of the Tetrahedron

$$Volume_{\text{tetrahedron}} = \frac{1}{3}\left(area_{\text{BCD}}\right)(h)$$

$$= \frac{1}{3}\left(\frac{\sqrt{3}}{4}ts^2\right)\left(\frac{\sqrt{2}}{\sqrt{3}}ts\right)$$

$$= \frac{\sqrt{2}}{12}ts^3$$

$$= \frac{1}{6\sqrt{2}}ts^3 = 0.11785113\,ts^3 \tag{1.2}$$

1.3. Surface Area of Tetrahedron

What is the surface area of the tetrahedron? It is just the sum of the areas of its 4 faces.

We know from above that the area of a face is $\frac{\sqrt{3}}{4}ts^2$.

The total surface area of the tetrahedron is

$$4\left(\frac{\sqrt{3}}{4}ts^2\right) = \sqrt{3}\,ts^2. \tag{1.3}$$

1.4. Side of Tetrahedron to Radius of Enclosing Sphere

Since all 4 vertices of the tetrahedron will fit inside a sphere, what is the relationship of the side of the tetrahedron to the radius of the enclosing sphere?

It's easier to see the radius of the enclosing sphere if we place the tetrahedron inside a cube (see Figure 1.3 below).

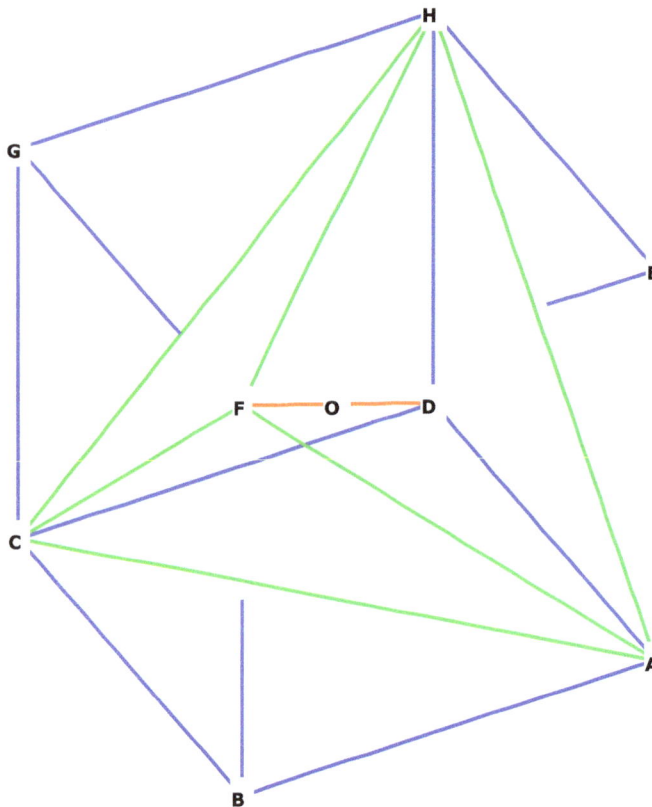

Figure 1.3: Tetrahedron in cube

The 4 vertices of the tetrahedron are H, F, C, A.

The 4 faces of the tetrahedron in this picture are: CFH, CFA, HFA, and at the back, HCA. (The vertex D is a part of the cube, not the tetrahedron).

The vertices of the cube all touch the surface of the sphere. The diameter of the sphere is the diagonal of the cube, $\overline{FD}$.

In this analysis and the ones following, we will assume that all of our polyhedra are enclosed within a sphere of radius 1. That way we will be able to accurately assess the relationship between all of the 5 regular solids.

The radius of this sphere is $\overline{OF} = \overline{OD} = 1$. How does the side of the tetrahedron relate to the radius of the sphere?

The side of the tetrahedron is the diagonal of the cube face, as can be seen in Figure 1.1 and Figure 1.3.

$\overline{FC}$ is the side of the tetrahedron, $\overline{DC}$ is the side of the cube, $\overline{FD}$ is the diagonal of the cube and the diameter of the sphere enclosing the cube and the tetrahedron.

By the Pythagorean Theorem, $\overline{FD}^2 = \overline{FC}^2 + \overline{DC}^2$.

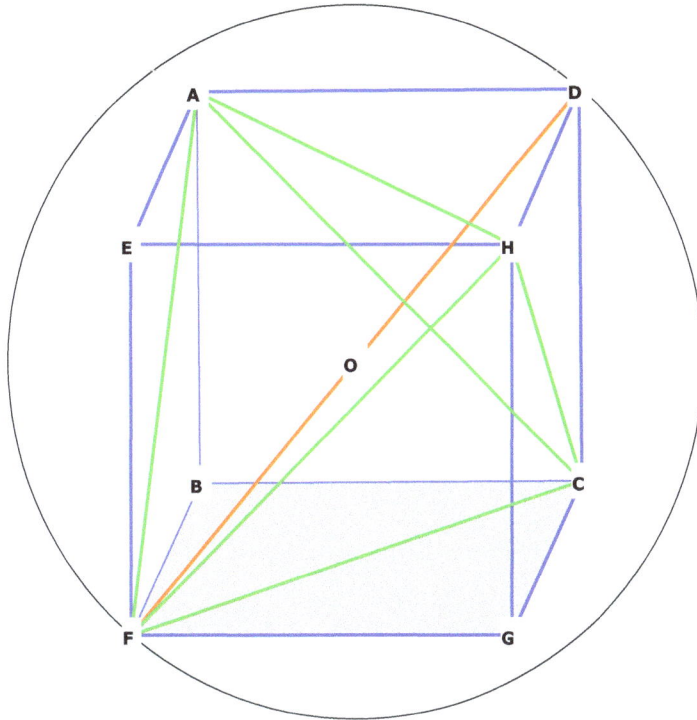

Figure 1.4: Cube and tetrahedron in unit sphere

$\overline{FC} = \sqrt{2}\left(\overline{FG}\right), \overline{DC} = 1.$

$\overline{FD}^2 = 2 + 1 = 3,$

$\overline{FD} = \sqrt{3}s$, where s = side of cube, and $\overline{OF} = \frac{\sqrt{3}}{2}s.$

So $\overline{FG}$ (side of cube), $\overline{FC}$ (side of tetrahedron), and $\overline{FD}$ (diameter of enclosing sphere) and we have a $1, \sqrt{2}, \sqrt{3}$ relationship.

1.4.1. RELATIONSHIP BETWEEN THE SIDE OF THE TETRAHEDRON AND THE RADIUS OF THE ENCLOSING SPHERE

We want to know the relationship between $\overline{OF}$ and $\overline{FC}$.

$\frac{\overline{FC}}{\overline{FD}} = \frac{\sqrt{2}}{\sqrt{3}}.$

$\overline{FD} = 2\left(\overline{OF}\right)$, so $\frac{\overline{FC}}{\overline{OF}} = \frac{2\sqrt{2}}{\sqrt{3}}.$

Therefore, $\frac{ts}{r} = \frac{2\sqrt{2}}{\sqrt{3}} = 1.632993162.$

So we write

$$ts = \frac{2\sqrt{2}}{\sqrt{3}}\, r, \, r = \frac{\sqrt{3}}{2\sqrt{2}}\, ts. \qquad (1.4)$$

1.5. Center of Mass of Tetrahedron

Where is the center of mass of the tetrahedron? We have marked it as O in the preceding figures. The centroid is just the distance O to any of the tetrahedron vertices, or the radius of the enclosing sphere. The center of mass of the cube is also the center of mass of the tetrahedron. As is evident from Figure 3 and Figure 7, every vertex of the tetrahedron is a vertex of the cube.

We already have $\overline{OF}$, or r, which is $\frac{\sqrt{3}}{2\sqrt{2}}\,ts$.

If we compare this distance to the height of the tetrahedron, we get

$$\frac{r}{h} = \frac{\frac{\sqrt{3}}{2\sqrt{2}}ts}{\frac{\sqrt{2}}{\sqrt{3}}ts} = \frac{\sqrt{3}}{2\sqrt{2}}\left(\frac{\sqrt{3}}{\sqrt{2}}\right) = \frac{3}{4} = 0.75.$$

Therefore,

$$\overline{AO} = 3\left(\overline{OH}\right) \tag{1.5}$$

.

Compare this to the equilateral triangle, where the relationship is 2 to 1.

1.6. Central Angle of Tetrahedron

What is the central angle of the tetrahedron? The central angle is the angle from one vertex, through the centroid O, to another vertex.

We calculate everything in terms of the side of the tetrahedron.

$\overline{OF} = \overline{OH} =$ radius of enclosing sphere.

$\overline{FH} = ts.$

$\overline{IF} = \frac{1}{2}\overline{FH} = \frac{1}{2}ts.$

$$\sin \angle IOF = \frac{\overline{OF}}{\overline{IF}} = \frac{1}{2}\left(\frac{2\sqrt{2}}{\sqrt{3}}\right) = \sin\left(\frac{\sqrt{2}}{\sqrt{3}}\right)$$

$$\angle IOF = \arcsin\left(\frac{\sqrt{2}}{\sqrt{3}}\right) = 54.73561032°$$

The central angle $\angle FOH = 2\left(\angle IOF\right).$

$$\text{Central angle} = 109.47122064° \tag{1.6}$$

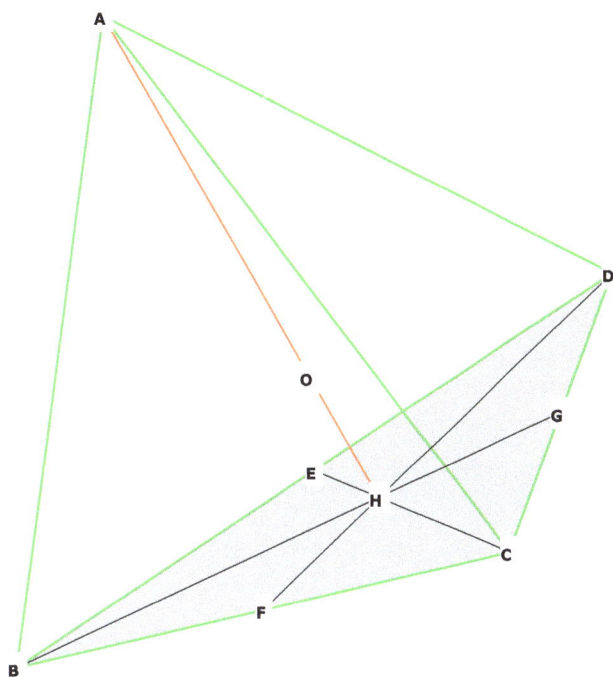

Figure 1.5: Figure 1.3 repeated

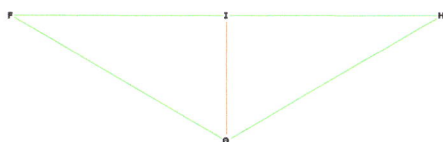

Figure 1.6: Central angle of tetrahedron

1.7. Surface Angle of Tetrahedron

What is the surface angle of the tetrahedron?

The surface angle is that angle made by the angles within the faces. Since each face of the tetrahedron is an equilateral triangle,

$$\text{Surface angle} = 60°. \tag{1.7}$$

1.8. Dihedral Angle of Tetrahedron

What is the dihedral angle of the tetrahedron?

The dihedral angle is the angle formed by the intersection of 2 planes:

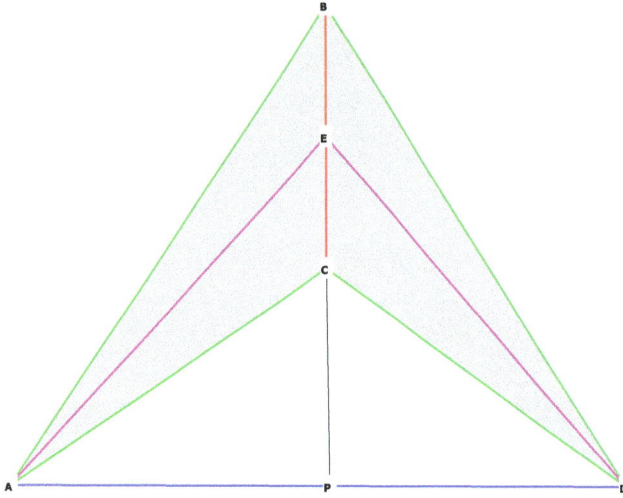

Figure 1.7: Tetrahedron dihedral angle

BAC and BDC are 2 intersecting faces of the tetrahedron. E is at the midpoint of $\overline{BC}$.

$\overline{AE}$ and $\overline{DE}$ are lines that go through the middle of each face and hit E.

The dihedral angle is AED.

The triangle $\triangle EPD$ is right. $\triangle PED = \frac{1}{2} \cdot \triangle AED$ by construction.

$\overline{AD}$ is the side of the tetrahedron s, so $\overline{PD} = \frac{1}{2}ts$.

$\overline{ED}$, the height of the tetrahedron face $= \frac{\sqrt{3}}{\sqrt{2}}ts$.

$$\sin\left(\angle PED\right) = \frac{\overline{PD}}{\overline{ED}} = \frac{\frac{1}{2}}{\frac{\sqrt{3}}{2}} = \frac{1}{2}\left(\frac{2}{\sqrt{3}}\right) = \frac{1}{\sqrt{3}}$$

$$\angle PED = \arcsin\left(\frac{1}{\sqrt{3}}\right) = 35.26438968°.$$

$$\angle AED = 2\left(\angle PED\right), \text{Dihedral angle} = 70.52877936°. \tag{1.8}$$

1.9. Centroid Distances

Now we will calculate the distances between the centroid and

- the middle of one of the faces $(\overline{OJ})$
- the midpoint of one of the sides $(\overline{OI})$
- any vertex $(\overline{OH})$

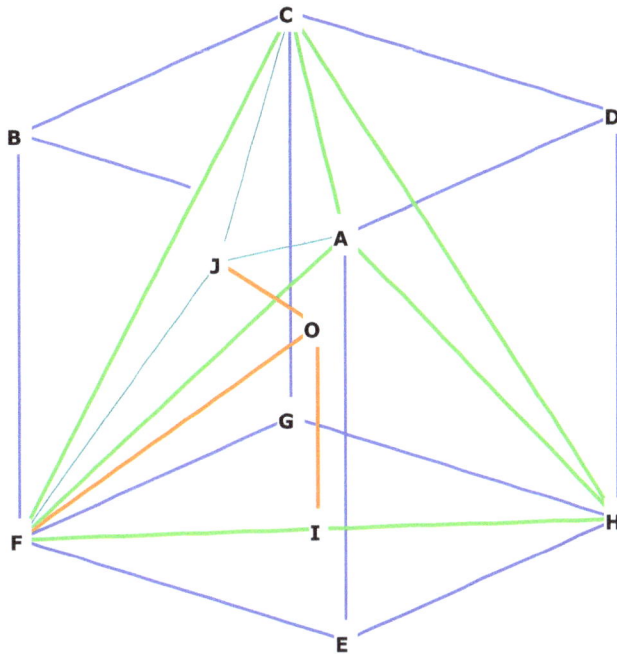

Figure 1.8: Centroid distances in Tetrahedron

$\overline{OF}$ = distance from centroid to a vertex, in this case, F,

$\overline{OI}$ = distance from centroid to middle of a side, in this case, the side $\overline{FH}$

$\overline{OJ}$ = distance from centroid to middle of a face, in this case, face ACF,

The 4 faces are CFH, CFA, HFA and HCA as before. The middle of face CFA is J. The midpoint of the side of the tetra $\overline{FH}$ is at I.

We have already figured out $\overline{OF} = \overline{OC} = \overline{OA} = \overline{OH}$, the distance from the centroid to any vertex. This is, as you recall, $\frac{\sqrt{3}}{2\sqrt{2}} ts$.

Let's get $\overline{OI}$ first.

$$\overline{OI}^2 = \overline{OF}^2 - \overline{IF}^2 = \left(\frac{\sqrt{3}}{2\sqrt{2}} ts\right)^2 - \left(\frac{1}{2} ts\right)^2 = \frac{3}{8} ts - \frac{1}{4} ts = \frac{1}{8} ts.$$

$$OI = \frac{1}{2\sqrt{2}} ts.$$

The triangle $\triangle FJO$ in Figure 1.8 is right because $\angle OJF$ is right.[1]

---

[1] Note to teachers: if you build an octahedron and a rhombic dodecahedron off the octahedron faces (use the Zometool), and then attach a tetrahedron off one of the octahedron faces, the centroid of that tetrahedron will be the vertex of the rhombic dodecahedron. In fact, the distance from the centroid of the tetrahedron to each of it's vertices is the edge length of the rhombic dodecahedron! Nature is not only beautiful, but also logical and consistent.

This will enable us to get the distance $\overline{OJ}$.

From Appendix B, *The Equilateral Triangle*, we know that $\overline{FJ} = \frac{1}{\sqrt{3}} ts$.

$$\overline{OJ}^2 = \overline{OF}^2 - \overline{FJ}^2$$

$$= \left(\frac{\sqrt{3}}{2\sqrt{2}} ts \right)^2 - \left(\frac{1}{\sqrt{3}} ts \right)^2$$

$$= \frac{3}{8} ts^2 - \frac{1}{3} ts^2 = \frac{1}{24} ts^2,$$

$$\overline{OJ} = \frac{1}{2\sqrt{6}} ts. \tag{1.9}$$

To get a good idea of relative distances, rewrite $\overline{OF}$ as $\frac{3}{2\sqrt{6}} ts$, and $\overline{OI}$ as $\frac{\sqrt{3}}{2\sqrt{6}} ts$.

Then the relationship between $\overline{OJ}, \overline{OI}$ and $\overline{OF}$ is $1, \sqrt{3}, 3$.

1.10. Tetrahedron Reference Tables

Table 1.1: Volume and Surface Area

Volume in terms of s	Volume in Unit Sphere	Surface Area in terms of s	Surface Area in Unit Sphere
0.11785113 s³	0.513200238 r³	1.732050808 s²	4.618802155 r²

Table 1.2: Angles

Central Angle:	Dihedral Angle:	Surface Angle:
109.47122064°	70.52877936°	60°

Table 1.3: Centroid Distances

Centroid To Vertex:	Centroid To Mid-edge:	Centroid To Mid-face:
1.0 r	0.577350269 r	0.33333333 r
0.612372436 s	0.353553391 s	0.204124145 s

Table 1.4: Side to Radius

Side / radius
1.632993162

Chapter 2

The Octahedron

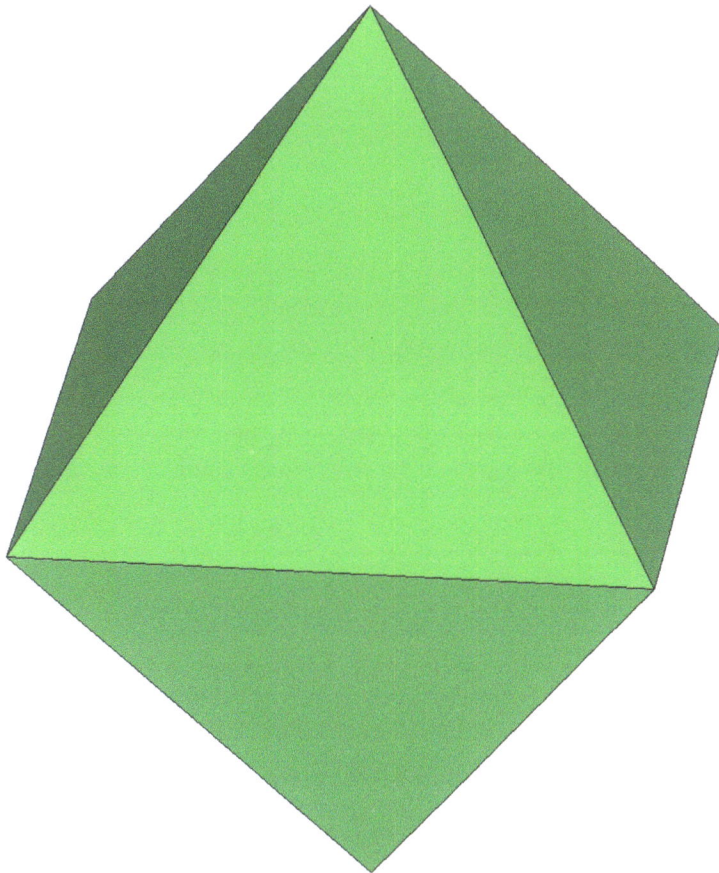

Figure 2.1: The Octahedron

The Octahedron has 12 sides, 8 faces and 6 vertices. Count them!

Each of the octahedron's 8 faces is an equilateral triangle, just like the tetrahedron, but the tetrahedron only has 4 faces.

Notice how the octahedron can be considered to be formed from 3 orthogonal squares:

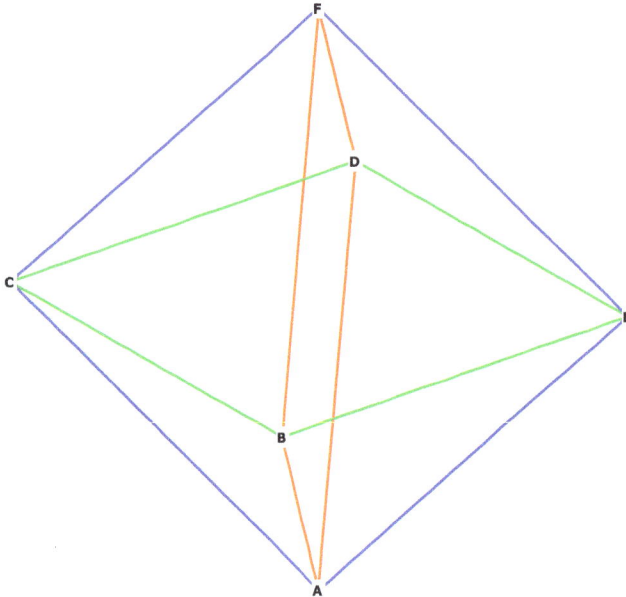

Figure 2.2: Octahedron, internal, 8 faces, 3 squares

The square BCDE, the square ABFD, and the square ACFE, all three of which are planes and all three of which are perpendicular to each other.

The octahedron's face angles are all equal because they are all 60°, however, the internal angles of the squares are all 90°. I guess it depends on which way you want to look at it! That's spatial geometry: the position of the observer relative to an object can yield quite different perspectives.

Let's calculate the volume of the octahedron. All of the octahedron's vertices touch upon the surface of the same sphere that encloses the tetrahedron. The centroid is at O. We will place a midpoint G in the middle of the face CDF. We also place a midpoint H on $\overline{CD}$, one of the sides of the octahedron. The side or edge of the octahedron will be hereinafter referred to as os.

First let's find the volume:

2.1. Volume of Octahedron

The octahedron consists of 2 pyramids, face-bonded. One pyramid is at A-BCDE, the other at F-BCDE.

The base of each pyramid is the square BCDE. The area of the base is then just $os \cdot os = os^2$.

The height of the pyramids are $h = \overline{OF} = \overline{OA}$.

But $\overline{OF} = \overline{OA} = \overline{OC}$, since the octahedron is composed of 3 squares. $\overline{FC}$ is just a side of the octahedron.

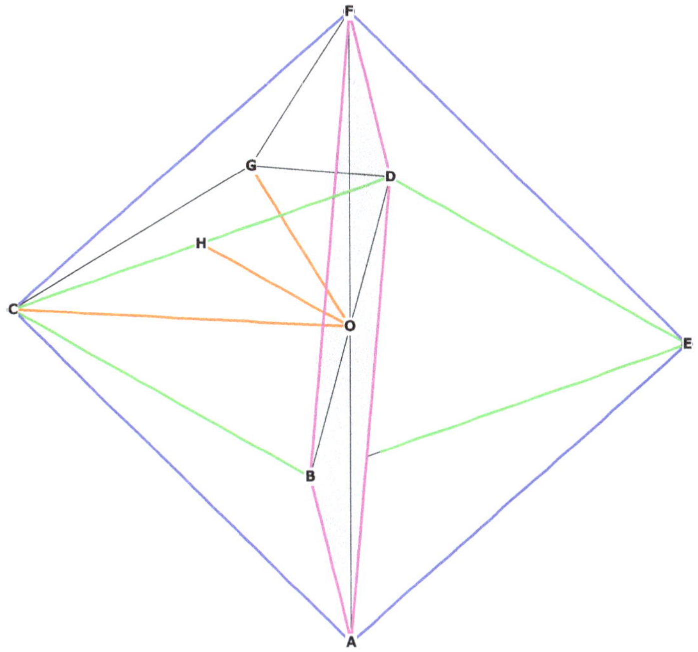

Figure 2.3: Inside the octahedron

Therefore we can write:

$$\overline{OF}^2 + \overline{OC}^2 = \overline{FC}^2,$$
$$\overline{OF}^2 + \overline{OF}^2 = \overline{FC}^2,$$
$$2\left(\overline{OF}^2\right) = os^2,$$
$$\overline{OF} = h = \frac{1}{\sqrt{2}}\,os.$$

$$Volume_{\text{1 pyramid}} = \frac{1}{3}\,(\text{area of base})\,(\text{height})$$

$$= \left(\frac{1}{3}\,os^2\right)\left(\frac{1}{\sqrt{2}}\,os\right),$$

$$= \frac{1}{3\sqrt{2}}\,os^3.$$

$$Volume_{\text{octahedron}} = 2\,(V_{\text{1 pyramid}}) = \frac{2}{3\sqrt{2}}\,os^3,$$

$$Volume_{\text{octahedron}} = \frac{\sqrt{2}}{3}\,os^3 = 0.471404521\,os^3. \tag{2.1}$$

2.2. Surface Area of Octahedron

What is the surface area of the octahedron? It is just the sum of the area of the faces. Since each face is an equilateral triangle, we know from *The Equilateral Triangle* that the area for one face is $\frac{\sqrt{3}}{4} os^2$.

$$Surface\ area\ _{octahedron} = 8 \left(\frac{\sqrt{3}}{4} os^2 \right) = 2\sqrt{3}\ os^2. \tag{2.2}$$

2.3. Octahedron Side to Radius of Enclosing Sphere

What is the relationship between the radius of the enclosing sphere and the side of the octahedron?

All 6 vertices of the octahedron touch the surface of the sphere. Therefore the diameter of the sphere is just $\overline{FA} = \overline{CE} = \overline{DB}$.

The radius is one-half that, or h, which we found above to be $\frac{1}{\sqrt{2}} os$.

So

$$r = \frac{1}{\sqrt{2}} os.\ os = \sqrt{2}\,r. \tag{2.3}$$

2.4. Central Angle of Octahedron

The central angle of the octahedron is $90°$.

The centroid of the octahedron is at O.

Since the octahedron is made of 3 orthogonal squares, the angle through O from any 2 adjacent vertices must be $90°$. To see this, look at $\angle FOC$.

2.5. Surface Angles of Octahedron

What are the surface angles of the octahedron? $60°$, because each of the faces is an equilateral triangle.

2.6. Dihedral Angle of Octahedron

What is the dihedral angle of the octahedron?

The dihedral angle is the angle formed by the intersection of 2 planes:

$\angle AXF$ is the dihedral angle because it is the intersection of the planes ABC and FBC.

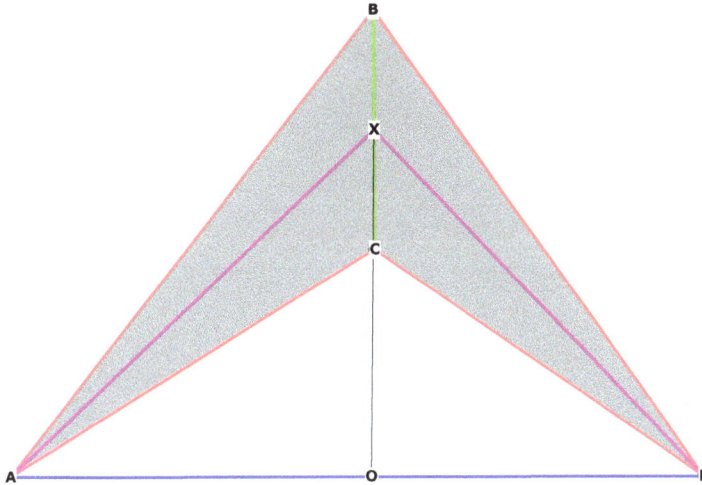

Figure 2.4: Octahedron dihedral angle

ACB and FBC are faces of the octahedron. $\overline{AX}$ and $\overline{XD}$ are lines from the face vertex through the mid-face and bisecting an edge of the octahedron, $\overline{BC}$.

$\overline{AF}$ is the diameter of the enclosing sphere, and $\overline{AO}$ and $\overline{OF} = r = h$.

OX is just one-half of a side of the octahedron $\overline{BC}$. You can see this marked in Figure 2.3 as the line $\overline{OH}$.

So $\overline{OX} = \frac{1}{2}\,os$.

Triangle $\triangle$XOF is right.

$$\tan{(OXF)} = \frac{OF}{OX} = \frac{\frac{1}{\sqrt{2}}}{\frac{1}{2}} = \frac{2}{\sqrt{2}} = \sqrt{2}.$$

$$\angle OXF = \arctan{(\sqrt{2})} = 54.73561032° = \frac{1}{2}(\angle AXF)$$

$$\text{Dihedral angle} = 109.4712206°. \qquad (2.4)$$

2.7. Centroid Distances

What is the distance from the centroid to the midpoint of one of the sides?

What is the distance from the centroid to the middle of one of the faces?

The distance from the centroid to the midpoint of a side can be seen in Figure 2.3 as $\overline{OH}$.

We already know this to be $\frac{1}{2}\,os$.

The distance from the centroid to the middle of a face can be seen in Figure 2.3 as $\overline{OG}$.

Triangle $\triangle FGO$ (see Figure 2.3) is right, because $\overline{OG}$ is perpendicular to the face FBC by construction.

From *The Equilateral Triangle* we know that $\overline{FG} = \frac{1}{\sqrt{3}} os$.

$$\overline{OF} = h = \frac{1}{\sqrt{2}} os.$$

$$\overline{OG}^2 = \overline{OF}^2 - \overline{FG}^2 = \frac{1}{2} os^2 - \frac{1}{3} os^2 = \frac{1}{6} os^2.$$

$$\overline{OG} = \frac{1}{\sqrt{6}} os.$$

The distance from the centroid O to the middle of one of the faces is $\frac{1}{\sqrt{6}}$ os.

What is the distance from the centroid to any vertex? It is just the radius of the enclosing sphere, or $\frac{1}{\sqrt{2}} os$.

2.8. Octahedron Reference Tables

Table 2.1: Volume and Surface Area

Volume in terms of s	Volume in Unit Sphere	Surface Area in terms of s	Surface Area in Unit Sphere
0.471404521 s³	1.33 r³	3.464101615 s²	6.92820323 r²

Table 2.2: Angles

Central Angle:	Dihedral Angle:	Surface Angle:
90°	109.4712206°	60°

Table 2.3: Centroid Distances

Centroid To Vertex:	Centroid To Mid-edge:	Centroid To Mid-face:
1.0 r	0.707106781 r	0.577350269 r
0.707106781 s	0.5 s	0.408248291 s

Table 2.4: Side to Radius

Side / radius
1.414213562

Chapter 3

The Cube

The cube has 12 sides, 8 vertices and 6 faces.

All of the vertices of the cube lie on the surface of the unit sphere. So the diameter of the enclosing sphere is $\overline{CE}$, or any diagonal of the cube that goes through the centroid O (see Figure 3.1).

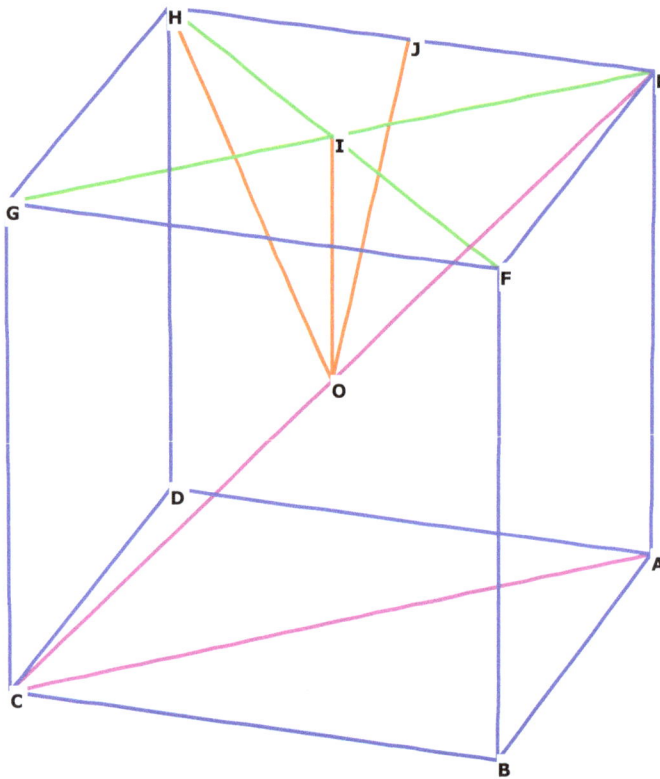

Figure 3.1: Cube internals

The radius of the sphere is one-half that; $\overline{OE}$, for example.

In other words, the radius is any line from O to any vertex.

The side of the cube will be referred to as s.

3.1. Volume of Cube

The volume of the cube is just $s \cdot s \cdot s = s^3$.

3.2. Surface Area of Cube

The surface area is just 6 faces multiplied by the area of each face.

The area of each face is $s \cdot s = s^2$, so the surface area is $6s^2$.

3.3. The Cube in the Unit Sphere

What is the relationship between the radius of the enclosing sphere and the side of the cube?

Look at the right triangle $\triangle OIE$ in Figure 3.1. To get the radius $\overline{OE}$ in terms of the side of the cube, notice that $\overline{OI}$ is just one-half $\overline{BF}$, the side of the cube.

So $OI = \frac{1}{2}s$.

$\overline{IE}$ is one-half the diagonal of the face EFGH. By the theorem of Pythagoras, $\overline{GE}$ is $\sqrt{2}\,s$, because $\overline{GH}$ and $\overline{HE}$ are both s.

Therefore, $\overline{IE} = \frac{1}{\sqrt{2}}s$.

$$\overline{OE}^2 = r^2 = \overline{OI}^2 + \overline{IE}^2 = \frac{1}{4}s^2 + \frac{1}{2}s^2 = \frac{3}{4}s^2.$$

$$\overline{OE} = \frac{\sqrt{3}}{2}s$$

$$r = \frac{\sqrt{3}}{2}s, \quad s = \frac{2}{\sqrt{3}}r. \tag{3.1}$$

The volume of the cube in the unit sphere of radius $= 1$ is therefore

$$s \cdot s \cdot s = \left(\frac{2}{\sqrt{3}}r\right)^3 = \frac{8}{3\sqrt{3}}r^3. \tag{3.2}$$

3.4. Central Angle of the Cube

What is the central angle of the cube?

The central angle is the angle through O from any two adjacent vertices. So in Figure 3.2, that would be HOE.

We bisect $\overline{HE}$ at J to get the triangle $\triangle OJE$, which is right by construction.

$\angle JOE$ will be one-half the central angle $\angle HOE$.

$$\sin(\angle JOE) = JE/OE = \frac{\frac{1}{2}}{\frac{\sqrt{3}}{2}} = \frac{1}{\sqrt{3}}.$$

$$\angle JOE = \arcsin\left(\frac{1}{\sqrt{3}}\right) = 35.26438968°.$$

$$\angle HOE = 2\,(\angle JOE) = 70.52877936°.$$

$$\text{Central angle} = 70.52877936°. \tag{3.3}$$

$$\text{Surface angle} = 90°. \tag{3.4}$$

$$\text{Dihedral angle} = 90°. \tag{3.5}$$

You can see this by observing the intersection of the planes BCGF and DCGH.

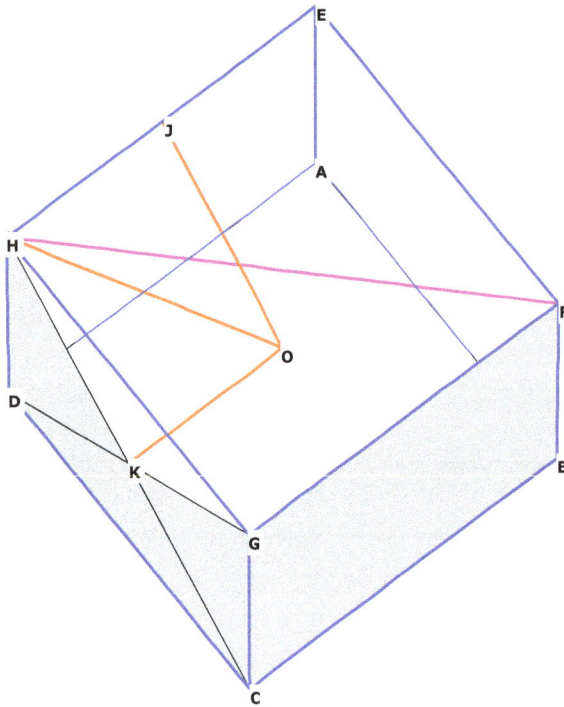

Figure 3.2: Dihedral angle of cube

3.5. Centroid Distances

The distance from the centroid to mid-face is $OI = \frac{1}{2}s$

The distance from the centroid to mid-edge is $\overline{OJ}$. Using the Pythagorean Theorem we may write

$$\overline{OJ}^2 = \overline{OE}^2 - \overline{JE}^2 = \left(\frac{\sqrt{3}}{2}s\right)^2 - \frac{1}{4}s^2 = \frac{3}{4}s^2 - \frac{1}{4}s^2 = \frac{1}{2}s^2.$$

$$\overline{OJ} = \frac{1}{\sqrt{2}}\,s. \tag{3.6}$$

The distance from the centroid to any vertex is

$$\overline{OE} = \frac{\sqrt{3}}{2}\,s. \tag{3.7}$$

Looking at comparative distances, we have .5, 0.707106781, 0.866015404.

3.6. Cube Reference Tables

Table 3.1: Volume and Surface Area

Volume in terms of s	Volume in Unit Sphere	Surface Area in terms of s	Surface Area in Unit Sphere
1.0 s³	1.539600718 r³	6.0 s²	8.0 r²

Table 3.2: Angles

Central Angle:	Dihedral Angle:	Surface Angle:
70.52877936°	90.0°	90.0°

Table 3.3: Centroid Distances

Centroid To Vertex:	Centroid To Mid-edge:	Centroid To Mid-face:
1.0 r	0.816496581 r	0.577350269 r
0.866025404 s	0.707106781 s	0.5 s

Table 3.4: Side to Radius

Side / radius
1.154700538

Part II

The Phi Ratio and the Pentagon

Chapter 4

The Phi Ratio

In order to properly study the remaining Platonic Solids, it is necessary to learn about the Phi ratio. It turns out that the dodecahedron and the icosahedron are both based on the $\sqrt{5}$ and Phi.

4.1. Introduction to the Phi ratio

The Phi Ratio is a proportion. A proportion is a relationship between one quantity and another quantity. It is often very helpful to have something to compare a thing to, in order to establish familiarity with it. For instance, to say "that building is 100" really doesn't help us at all. But if we say "the building is 100 stories" we can get an idea in our minds how large it is. By associating a metric, or measurement, to a number, we give it meaning and make it real.

In geometry, comparing two quantities to each other helps us establish a relationship between the two quantities. A pure number has no reality, because there is nothing to compare that number to. In geometry, if we can start with a known quantity or length, and then compare every other element in the drawing to that known quantity, we can stay clear on how everything in the drawing relates to everything else.

The Phi Ratio has been known for millennia. It has also been referred to as the Golden Ratio. Euclid called it "Division in Mean and Extreme Ratio." I sometimes shorten this to EMR.

The division into mean and extreme ratio is extremely important because in such a division, there is perfection. The division of any line into EMR, for example, can be continued infinitely small or infinitely large without the slightest error, or "round off" of any kind. So division in EMR or Phi Ratio leads to perfect harmonious growth.

The Phi ratio comes from the division of a line segment such that, "The smaller is to the larger as the larger is to the whole."

Consider the line segment $\overline{GX}$, divided at O into line segments $\overline{GO}$ with smaller length a, and $\overline{OX}$, with larger length b.

Figure 4.1: Line divided into Mean and Extreme ratio

Mathematically stated, the above statement becomes

$$\frac{a}{b} = \frac{b}{a+b}.$$

Let $b = \overline{OX} = 1$. Then

$$\frac{a}{1} = \frac{1}{a+1}.$$

And

$$a^2 + a = 1 \quad \text{and} \quad a^2 + a - 1 = 0.$$

$a^2 + a = 1$ is a second degree polynomial, and you can solve it easily on your calculator. Just press the "poly" key and enter $1, 1, -1$.

The calculator will show two solutions, the first is

$$0.618033989 = \frac{1}{\Phi}.$$

The second solution is

$$-1.618033989 = -\Phi.$$

Because a is shorter than b, and a is a positive length, a must be $\frac{1}{\Phi}$. Then

$$a + b = 1 + \frac{1}{\Phi} = \Phi.$$

"The smaller is to the larger:" The ratio

$$\frac{b}{a} = \frac{1}{\frac{1}{\Phi}} = \Phi. \tag{4.1}$$

"As the larger is to the whole:"

$$\frac{b+a}{b} = \frac{\Phi}{1} = \Phi. \tag{4.2}$$

Although Φ is a number it is more properly stated as a geometric ratio of two numbers. A ratio is a relationship between two things. When we investigate the pentagon, we'll see that Phi is stated mathematically as $\left(\frac{\sqrt{5}+1}{2}\right)$, and that $\frac{1}{\Phi}$ is stated mathematically as $\left(\frac{\sqrt{5}-1}{2}\right)$. This might seem uninformative at first, but there is a very understandable and intuitively sensible geometric interpretation of these quantities that you'll be able to easily grasp.

Now let's look at the Phi Ratio in action.

4.2. The Golden Triangle

The golden triangle is a phi ratio triangle:

Triangle $\triangle ABC$ is a golden triangle because when either of the long sides ($\overline{AB}$ or $\overline{AC}$) is divided by the short side ($\overline{BC}$), the result is Phi (Φ).

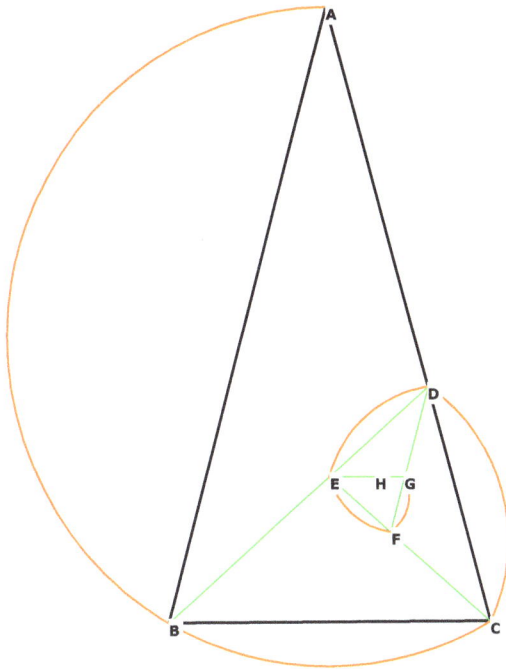

Figure 4.2: Golden triangle

You can see how the triangle divides into itself, always making a triangle with sides and angles in the same proportion. For example, the triangle $\triangle BCD$ is also a golden triangle. Triangles CDE, DEF, and EFG are as well, all having the property that the long side of each triangle divided by the short side $= \Phi$.

In the large triangle $\triangle ABC$, for example, if the line segments $\overline{AB}$ and $\overline{BC}$ were combined to form one line, that line would be divided in Phi ratio at B:

Figure 4.3: The line ABC divided in Phi Ratio at B.

4.2.1. EXERCISE:

Requires paper, compass and straight-edge. Take the smaller segment of triangle $\triangle ABC$, $\overline{BC}$, and transfer that length to the line $\overline{AC}$. Do this by pinning the compass at B with length BC to get the distance, then swinging the marker end until it hits $\overline{AC}$ at D. Now $\overline{AC}$ has been divided in Phi ratio at the point D! Connect $\overline{BD}$ with the straight-edge. A second triangle has been now formed, BCD, which is itself a golden mean triangle. The process is absurdly simple, for once a Golden Triangle is constructed, it can be subdivided or grown with hardly any thought whatsoever!

If we let $\overline{AB} = \overline{AC} = 1$, then

$\overline{BC} = \overline{BD} = \overline{AD} = \frac{1}{\Phi}$.

$DC = \frac{1}{\Phi^2}$, because $\frac{1}{\Phi} + \frac{1}{\Phi^2} = 1$.

Now do the same thing with triangle $\triangle BCD$. Take the short side $\overline{CD}$, and transfer that distance onto $\overline{BD}$ to find the point E. Then connect $\overline{CE}$ with the straight-edge. Another golden mean rectangle is formed, CDE. Continue in this fashion to subdivide the larger segment of each triangle by the smaller.

You can see that this process can occur indefinitely. The process can be reversed to make the object larger, on to infinity. The only reason this works is that the division into Phi ratio occurs absolutely perfectly, with not even the tiniest error. (Of course there is always construction error, but mathematically the Phi ratio is perfect division).

$$\overline{BC} = \overline{BD} = \overline{AD} = \frac{1}{\Phi},$$

$$\overline{CD} = \overline{CE} = \overline{BE} = \frac{1}{\Phi^2}.$$

$$\overline{DE} = \overline{DF} = \overline{CF} = \frac{1}{\Phi^3}.$$

$$\overline{EF} = \overline{EG} = \overline{DG} = \frac{1}{\Phi^4}... \tag{4.3}$$

A little trigonometry reveals that these golden triangles have angles 36–72–72. (Do this as an exercise).

Notice the spiral that forms around the triangles. This is called a logarithmic spiral. If you observe nature, you will find that she does not use straight lines, she uses curves.

4.3. The Golden Rectangle

See Figure 4.4 on p. 34. The long sides of the rectangle ABGF ($\overline{AB}$ and $\overline{FG}$) have first been divided into Phi ratio at E and H; meaning that, for example, the ratio $\frac{FG}{FH} = \Phi$, and the ratio $\frac{FH}{GH} = \Phi$. Remember that Phi is a ratio describing the relationship between two quantities.

4.3.1. EXERCISE:

To make the smaller golden rectangle EBGH inside ABGF, simply pin your compass at A and mark out the distance to F. Then pin your compass at F and mark off the distance to H on the line $\overline{FG}$. With the straightedge, draw a perpendicular to the point H. Now EBGH is another golden rectangle. Do the same to make golden rectangle EBIJ, and so on. Notice that AEHF is a square, and so is JIGH, and so on. This process can go on downward or upward indefinitely. Again, it is really simple and painless. Once you get the hang of it, perfect growth or division can occur almost automatically.

Notice that the spiral hits all of the points where a line has been divided into Phi ratio: H, where $\overline{FG}$ has been divided into Phi ratio; I, where $\overline{GB}$ has been divided into Phi ratio, etc.

What have we learned? That the division into Phi ratio leads to absolutely perfect, harmonious growth. Unfortunately, perfection has no beginning or ending. In other words, we could make our rectangles smaller and smaller, and they would begin to converge at a particular point, but never reach it. As you can see from the above diagram, the spiral keeps circling round and round, never quite reaching the center, or growing outward without end. When nature builds something, she needs a starting and an ending point. Fortunately, hidden within the properties of Phi, is the answer.

Next: *The Fibonacci Series.*

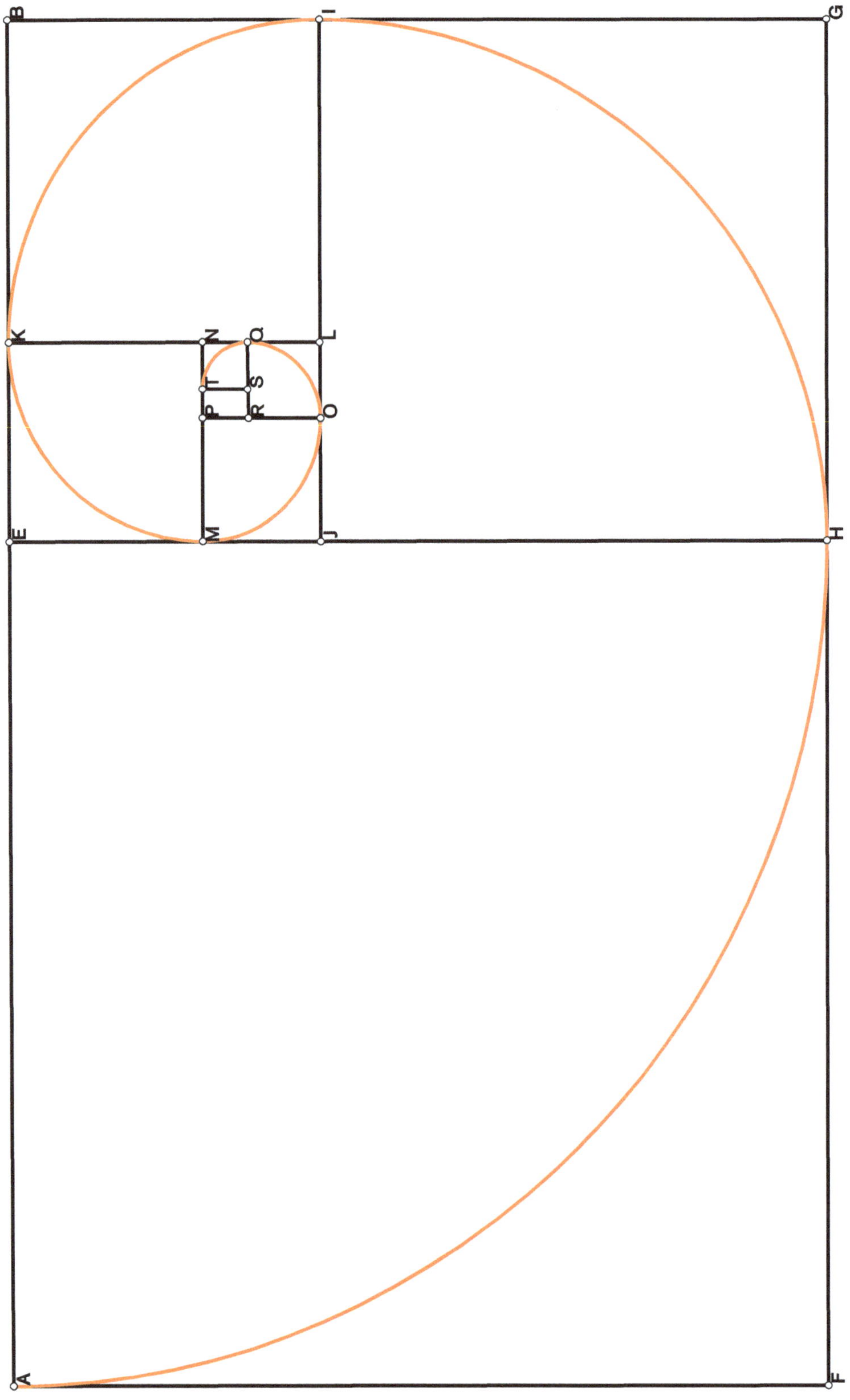

Figure 4.4: Golden rectangle

Chapter 5

The Fibonacci Series

Because division or growth by the Golden Mean has no beginning or ending, it is not a good candidate for constructing forms of any kind. It is necessary to have a definable starting point if you want to build anything.

However, there is an approximation to the Golden Mean that nature uses, called the Fibonacci Sequence. Leonardo Fibonacci was a monk who noticed that branches on trees, leaves on flowers, and seeds in pine cones and sunflower seeds arranged themselves in this sequence.

The Fibonacci Sequence is based on the golden mean ratio, or Phi (Φ). To the left of the equal sign is Φ raised to a power, and to the right of the equal sign is the Fibonacci Sequence:

$$
\begin{aligned}
\Phi^1 &= 1\Phi + 0, \\
\Phi^2 &= 1\Phi + 1, \\
\Phi^3 &= 2\Phi + 1, \\
\Phi^4 &= 3\Phi + 2, \\
\Phi^5 &= 5\Phi + 3, \\
\Phi^6 &= 8\Phi + 5, \\
\Phi^7 &= 13\Phi + 8, \\
\Phi^8 &= 21\Phi + 13, \\
\Phi^9 &= 34\Phi + 21, \\
\Phi^{10} &= 55\Phi + 34,
\end{aligned}
\tag{5.1}
$$

... and so on.

Each digit in the second column to the left of the Φ symbol is the sum of the two before it in the previous row:

$1 + 0 = 1, \ 1 + 1 = 2, \ 2 + 1 = 3, \ 3 + 2 = 5, \ 5 + 3 = 8, \dots$ and so on.

Notice that when one divides a digit by the one before it in the sequence, the result rapidly approaches Φ:

$$\frac{1}{1} = 1.0$$

$$\frac{2}{1} = 2.0$$

$$\frac{3}{2} = 1.5$$

$$\frac{5}{3} = 1.67$$

$$\frac{8}{5} = 1.60$$

$$\frac{13}{8} = 1.625$$

$$\frac{21}{13} = 1.6153846$$

$$\frac{34}{21} = 1.6190476$$

$$\frac{55}{34} = 1.6176471$$

Φ is, numerically, 1.618033989....

Here is a chart that shows this:

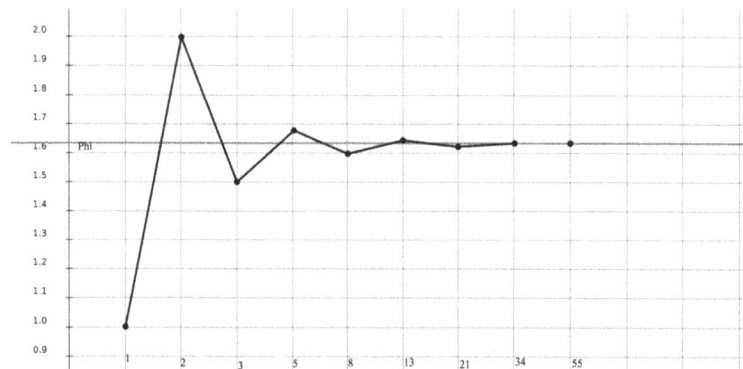

Figure 5.1: How the Phi ratio approaches Φ

Notice that Φ is approached both from above and below. This is not an asymptotic approach, but one which affords a full view of both sides. Like a signal locking in on the target, the Fibonacci sequence homes in on Φ.

The Φ ratio is attained from the division of integers. Integers, or whole numbers, represent things that can be designed and built in the "real world." Even though Φ is unattainable, nature can closely approximate it, as we can see from the following photographs:

Figure 5.2: Nautilus shell. Image taken from *Wikimedia Commons*. User: Chris **73**

Figure 5.3: Echeveria Agavoides (molded wax agave). Image taken from *Wikimedia Commons*, Echeveria agavoides in Kultur bei Kakteen-Haage in Erfurt, Michael Wolf.

Figure 5.4: Helianthus Annus (common sunflower). Author photograph

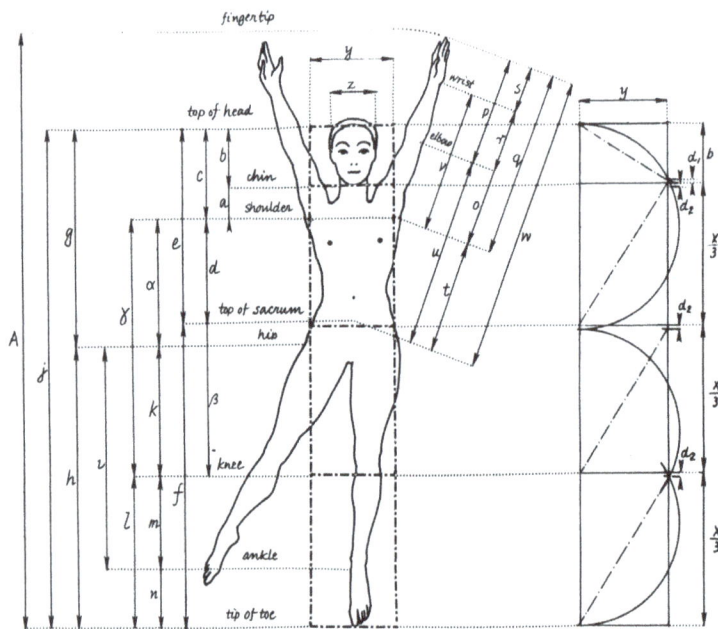

Figure 5.5: Human body phi relationships. Image taken from *The Power of Limits*, Gyorgy Doczi, p. 143

The human body has many Phi/Fibonacci relationships, as shown in the image above.

The leg: The distance from the hip to the knee, and from the knee to the ankle.

The face: The distance from the top of the head to the nose, and from the nose to the chin.

The arm: The distance between the shoulder joint to the elbow, and from the elbow to the wrist.

The hand: The distance between the wrist, the knuckles, the first and second joints of the fingers, and the finger tips.

Of course every body is different and these relationships are only approximate. Once you become aware of the Fibonacci sequence / Phi ratio, however, you begin to see it in many life forms.

(As an aside, note that in the Fibonacci sequence, you don't have to start with 1. Begin with any number and proceed with the rule "the current number is the sum of the two previous numbers." Do the division of the current number by the prior number in the sequence and Φ will always be the result. For example, begin with 5. Then the sequence goes 5, 5, 10, 15, 25, 40, 65, 105…The ratios between the two numbers will be exactly the same as the sequence progresses.)

On to *The Properties of Phi*

Properties of the Phi Ratio

Because Φ divides into itself perfectly, it has the following amazing properties:

$$\Phi^{-1} = \Phi^{-2} + \Phi^{-3}$$
$$\Phi^0 = \Phi^{-1} + \Phi^{-2}$$
$$\Phi^1 = \Phi^0 + \Phi^{-1}$$
$$\Phi^2 = \Phi^1 + \Phi^0$$
$$\Phi^3 = \Phi^2 + \Phi^1$$
$$\Phi^4 = \Phi^3 + \Phi^2...$$

So $\Phi^n = \Phi^{n-1} + \Phi^{n-2}$, n is any integer. Φ to any power is the sum of the two powers before it. Multiples of Φ can be combined to form any integer. Here is a brief table:

$$1 = \frac{1}{\Phi} + \frac{1}{\Phi^2} \text{ or } \Phi - \frac{1}{\Phi}$$
$$2 = \Phi + \frac{1}{\Phi^2} \text{ or } \Phi^2 - \frac{1}{\Phi}$$
$$3 = \Phi^2 + \frac{1}{\Phi^2}$$
$$4 = \Phi^3 - \frac{1}{\Phi^3} \text{ or } \Phi^2 + 1 + \frac{1}{\Phi^2}$$
$$5 = \Phi^3 + \frac{1}{\Phi} + \frac{1}{\Phi^4} \text{ or } 5\left(\Phi - \frac{1}{\Phi}\right)$$
$$6 = \text{multiple of } 2, 3$$
$$7 = \Phi^4 + \frac{1}{\Phi^4}$$
$$8 = \text{multiple of } 2, 4$$
$$9 = \text{multiple of } 3$$
$$10 = \text{multiple of } 2, 5$$
$$11 = \Phi^5 - \frac{1}{\Phi^5}, \text{ or } \Phi^4 + \Phi^2 + 1 + \frac{1}{\Phi^2} + \frac{1}{\Phi^4} \text{ (notice the symmetry)}$$
$$12 = \text{multiple of } 2, 3, 4$$
$$13 = \Phi^5 + \Phi + \frac{2}{\Phi^4}, \text{ and so on.}$$

Apparently, powers of Phi added or subtracted to their inverses always equals an integer.

Because Φ divides into itself perfectly, it has amazing powers of reduction. To see this mathematically, check out Appendix E, *Phi Reduction*.

On to *Construction of the Pentagon*

Chapter 7

The Pentagon, Part I

7.1. Constructing the Pentagon

In this section we'll learn how to construct the pentagon with compass and straight-edge. In the next section, we'll look at what we've done mathematically. Why are we spending so much time on the pentagon? Because it is fundamental to the construction of both the icosahedron and the dodecahedron. To follow along, get out your compass and straight-edge.

First, begin with a circle as usual. Pick a point anywhere on the circle, say X, and with the straightedge draw a line through O, the center, until it intersects the circle again at X'. The radius of the circle is $\overline{OX} = \overline{OX'}$. Let this distance be 1.

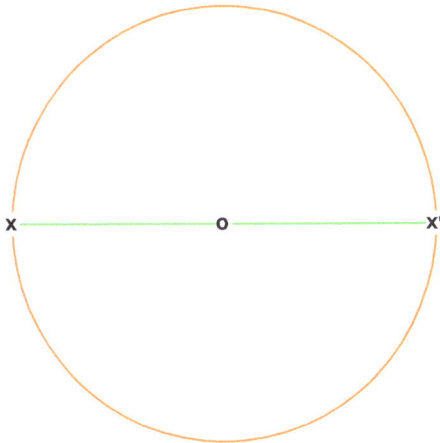

Figure 7.1: Construction of pentagon, step 1

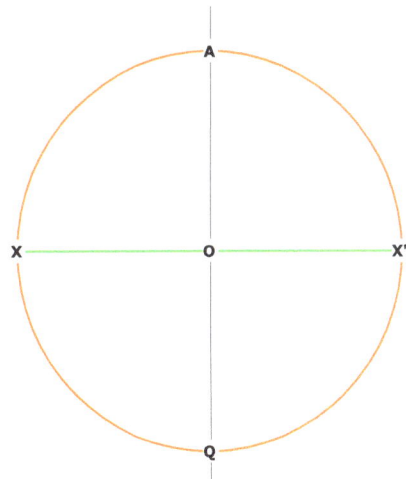

Figure 7.2: Construction of pentagon, step 2

Next, bisect the line $\overline{XX'}$ with the compass. Pin compass at X, place the other leg at X' and draw an arc from top (above A) to bottom (below Q).

Pin compass at X', and place the other leg at X. Draw an arc to intersect the arcs you just drew. I have placed two little x's on $\overline{AQ}$ where both arcs meet. Connect the x's and you have the line $\overline{AQ}$ in Figure 2:

Next, bisect the line $\overline{OX'}$ at P. We have now divided $\overline{OX'}$ in half. With the straight-edge connect line $\overline{AP}$.

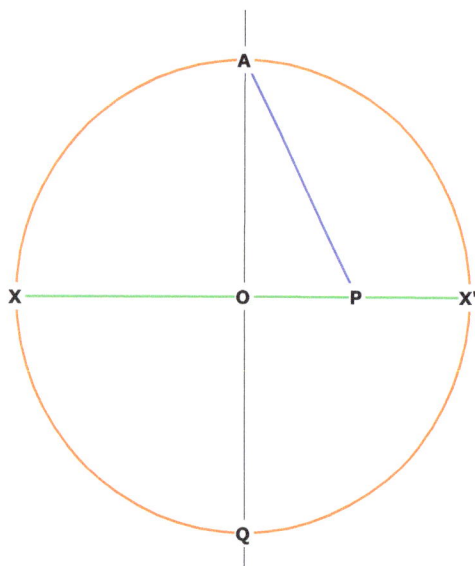

Figure 7.3: Construction of pentagon, step 3   Figure 7.4: Construction of pentagon, step 4

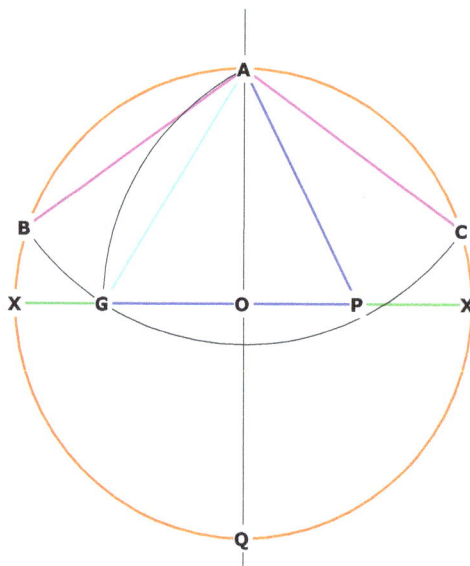

Now take the distance $\overline{AP}$ and transfer it to $\overline{XX'}$. Do this by pinning the compass at P, with length $\overline{AP}$, and drawing an arc down to $\overline{XX}$'. Mark that point as G. Now $\overline{AP} = \overline{PG}$ and the distance $\overline{AG}$ (marked in cyan) is the side of the pentagon. ($\overline{AG}$ is $>\overline{AP}$).

Now, pin the compass at A, with length $\overline{AG}$.

Draw an arc with the compass from $\overline{AG}$ through to B, and the other way until it hits the side of the circle at C. Now we have three of the 5 points of the pentagon, A, B, and C.

Now, walk the compass around the circle by either pinning it at B or C, with distance $\overline{AB} = \overline{AC}$, and finding D and E. Connect up ABDECA and you have the pentagon (see Figure 7.5 on next page).

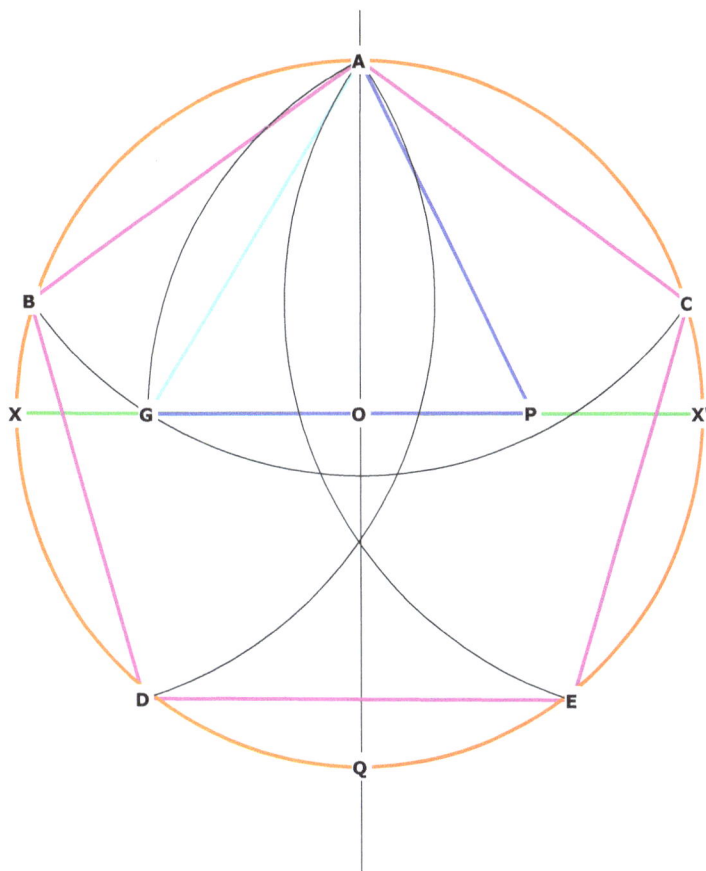

Figure 7.5: Constructing the pentagon, step 5

Why is the pentagon so important?

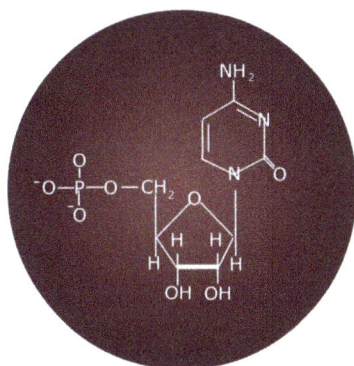

Figure 7.6: A nucleotide

Nucleotides are the basic units of DNA. The phosphate group to the left is square-shaped and the nitrogenous base is a hexagon. The central structure, the deoxyribose sugar, is pentagonally shaped. Nature uses geometry in the construction of life!

On to *Construction of the Pentagon, Part II*

Chapter 8

Construction of the Pentagon, Part II

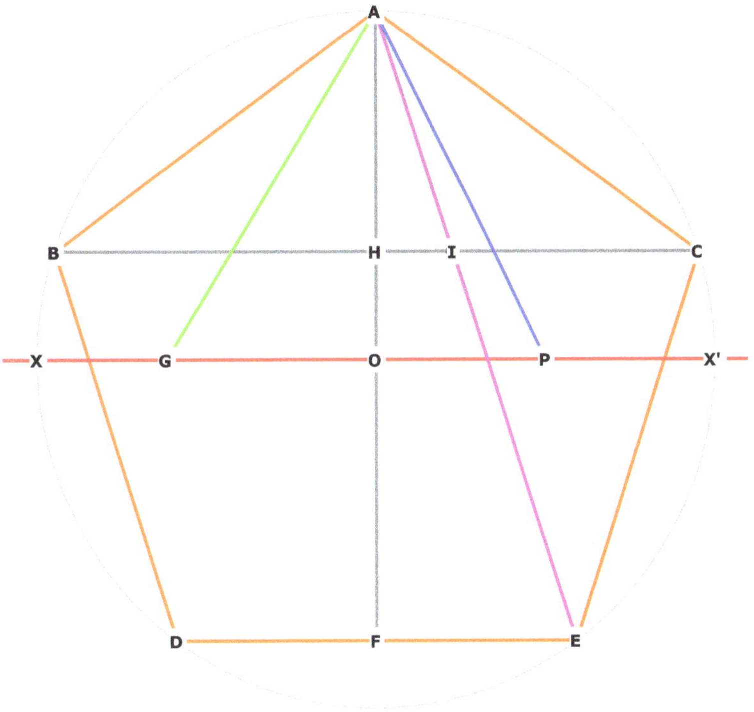

Figure 8.1: Pentagon internals

Something very important happens during the construction of the pentagon: the triangle $\triangle AOG$.

$\overline{AO} = \overline{OX}' = 1$, which is the radius.

We bisected $\overline{OX'}$ at P, so $\overline{OP} = \overline{PX'} = \frac{1}{2}$.

By the Pythagorean Theorem,

$$\overline{AP}^2 = \overline{AO}^2 + \overline{OP}^2 = 1 + \frac{1}{4} = \frac{5}{4}.$$

$$\overline{AP} = \frac{\sqrt{5}}{2} = \overline{GP}.$$

$$\overline{GX'} = \overline{GP} + \overline{PX'}, \text{so } \overline{GX'} = \frac{\sqrt{5}}{2} + \frac{1}{2}. \text{ This value is called Phi } (\Phi). \qquad (8.1)$$

$$\overline{GO} = \overline{GP} - \overline{OP}, \text{so } \overline{GO} = \frac{\sqrt{5}}{2} - \frac{1}{2}. \text{ This value is the inverse of Phi, or} \frac{1}{\Phi}. \qquad (8.2)$$

In Figure 8.1, the distance $\overline{AP}$ (the hypotenuse of triangle $\triangle AOP$)[1] is transferred to the line $\overline{XX'}$ at G, such that $\overline{AP} = \overline{PG}$. When this occurs, the line $\overline{GOX'}$ has been divided into Phi ratio at O.

Euclid called this the division into Mean and Extreme ratio.

The larger segment of $\overline{GX'}$ is $\overline{OX'}$, the smaller segment is $\overline{GO}$.

$\overline{GX'} = \Phi = 1.618033989...$

$\overline{GO} = \frac{1}{\Phi} = 0.618033989...$

The triangle $\triangle AOG$ has it's larger side $(\overline{OA}) = 1$ and its smaller side $(\overline{OG}) = \frac{1}{\Phi}$. Triangle $\triangle AOG$ is a Phi right triangle.

The hypotenuse of triangle $\triangle AOG$, $\overline{AG}$, is the result of the division of the line segments $\overline{GO}$ and $\overline{OA}$ into the Phi ratio.

8.1. What is $\overline{AG}$ in Figure 8.1?

By the Pythagorean theorem,

$$\overline{AG}^2 = \overline{OA}^2 + \overline{OG}^2 = 1 + \frac{1}{\Phi^2}.$$

---

[1]Phi occurs geometrically within a $1, \frac{1}{2}, \frac{\sqrt{5}}{2}$ triangle, when the short side distance is transferred to the hypotenuse. That distance divides the long side of the triangle in Phi ratio, like so:

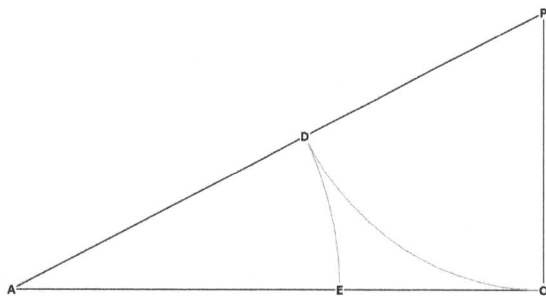

$\overline{OP}$ is transferred to $\overline{AP}$ at D, and then $\overline{AD}$ is transferred to $\overline{AO}$ at E. $\overline{AO}$ has been divided in Mean and Extreme Ratio at E. The triangle $\triangle AOP$ in Figure 8.1 is just such a triangle.

$$\overline{AG} = \sqrt{1 + \frac{1}{\Phi^2}}.$$

$\overline{AG} = \overline{AB}$, the side of the pentagon. This is the secret to the design of the pentagon – its sides are a result of a Phi ratio triangle. In the next section we will draw diagonals within the pentagon, and we will see that every single one of them intersects the others in phi ratio, forming smaller and smaller pentagons within each other, and forming many phi ratio triangles. The intersection of diagonals $\overline{AE}$ and $\overline{BC}$, for example, divide each other in Phi ratio at I, and form the Phi ratio triangle $\triangle AIC$. $\triangle ABC$ is also a phi ratio triangle, as is triangle $\triangle ACE$!

8.2. What is the relationship between the radius of the circle enclosing the pentagon, and the side of the pentagon?

We know that $\overline{AB} = $ side of pentagon(s) $= \left(\sqrt{1 + \frac{1}{\Phi^2}}\right) \cdot$ radius.

$\overline{OA} = \overline{OB} = \overline{OC} = \overline{OD} = \overline{OE} = $ radius.

So $r = \frac{1}{\sqrt{1 + \frac{1}{\Phi^2}}} s.$

$1 + \frac{1}{\Phi^2} = \frac{\Phi^2 + 1}{\Phi^2}$. So we rewrite

$$r = \frac{\Phi}{\sqrt{\Phi^2 + 1}} s = 0.850650808s. \tag{8.3}$$

8.3. What is the height of the pentagon AF?

$$\overline{AF}^2 = \overline{AE}^2 - \overline{FE}^2 = \Phi^2 s^2 - \frac{1}{4}s^2 = \frac{4\Phi^2 - 1}{4}s^2 = \frac{\Phi^2(\Phi^2 + 1)}{4}s^2.$$

So

$$h = \overline{AF} = \text{square root of that} = \frac{\Phi\sqrt{\Phi^2 + 1}}{2} s = 1.538841769\, s. \tag{8.4}$$

8.4. What is $\overline{AH}$?

$\overline{AH}$ bisects $\overline{BC}$, a diagonal of the pentagon. We know that the diagonal of a pentagon is (Φ) side of the pentagon. Therefore

$\overline{HC} = \frac{1}{2}(\Phi)(s) = \frac{\Phi}{2}s.$

$$\overline{AH}^2 = \overline{AC^2} - \overline{HC^2} = s^2 - \left(\frac{\Phi}{2}s\right)^2 = \frac{4 - \Phi^2}{4}s^2.$$

$$\overline{AH} = \frac{\sqrt{4 - \Phi^2}}{2} s = \frac{\sqrt{\Phi^2 + 1}}{2\Phi}s = 0.587785252\, s. \tag{8.5}$$

8.5. What is $\overline{OF}$?

$$\overline{OF} = \overline{AF} - \overline{AO}$$

$$= \frac{\Phi\sqrt{\Phi^2 + 1}}{2}s - \frac{\Phi}{\sqrt{\Phi^2 + 1}}s$$

$$= \frac{\Phi(\Phi^2 + 1) - 2\Phi}{2\sqrt{\Phi^2 + 1}}s$$

$$= \frac{\Phi(\Phi + 2) - 2\Phi}{2\sqrt{\Phi^2 + 1}}s$$

$$= \frac{\Phi^2}{2\sqrt{\Phi^2 + 1}}s = 0.688190961s. \tag{8.6}$$

8.6. What is $\overline{FH}$?

$$\overline{FH} = \overline{AF} - \overline{AH}$$

$$= \frac{\Phi\sqrt{\Phi^2 + 1}}{2}s - \frac{\sqrt{\Phi^2 + 1}}{2\Phi}s$$

$$= \frac{\Phi^2\sqrt{\Phi^2 + 1} - \sqrt{\Phi^2 + 1}}{2\Phi}$$

$$= \frac{\sqrt{\Phi^2 + 1}(\Phi^2 - 1)}{2\Phi}s$$

$$= \frac{\Phi\sqrt{\Phi^2 + 1}}{2\Phi} = \frac{\sqrt{\Phi^2 + 1}}{2} = 0.951056517\,s. \tag{8.7}$$

Note that :

$$\frac{\overline{FH}}{\overline{AH}} = \frac{\frac{\sqrt{\Phi^2 + 1}}{2}}{\frac{\sqrt{\Phi^2 + 1}}{2\Phi}} = \frac{\sqrt{\Phi^2 + 1}}{2}\left(\frac{2\Phi}{\sqrt{\Phi^2 + 1}}\right) = \Phi! \tag{8.8}$$

Therefore, the diagonal $\overline{BC}$ divides the height of the pentagon, $\overline{AF}$, in mean and extreme ratio.

8.7. What is the angle BAC?

We know that $\overline{AC} = 1$, and that $\overline{HC} = \frac{\Phi}{2}$, because $\overline{BC} = \Phi$. Therefore,

$$\sin\left(\angle HAC\right) = \frac{\overline{HC}}{\overline{CA}} = \frac{\Phi}{2},$$

$$\angle HCA = \arcsin\left(\frac{\Phi}{2}\right) = 54°$$

$\overline{FA}$ bisects $\angle BAC$, therefore

$$\angle BAC = 2\left(\angle HCA\right) = 108° \qquad (8.9)$$

8.8. The Area of the Pentagon

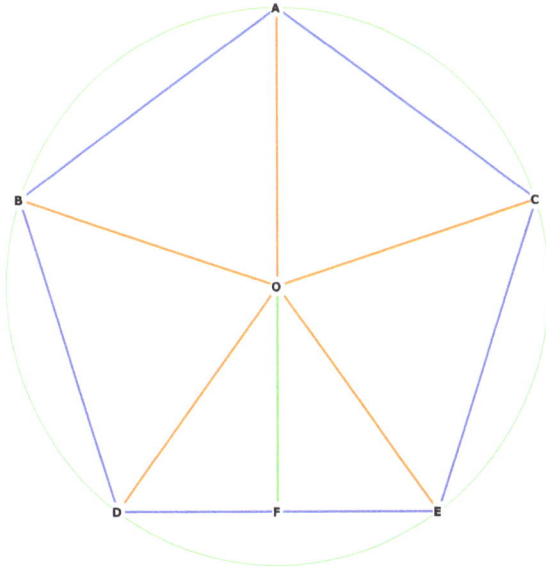

Figure 8.2: Area of pentagon

The area of the pentagon is the combined area of the 5 triangles shown in Figure 8.2, each of which has a vertex at O.

8.8.1. WHAT IS THE AREA OF EACH OF THESE TRIANGLES?

Lets work with triangle $\triangle ODE$.

From above we know that the height of triangle $\triangle ODE$, $\overline{OF}$, is $\frac{\Phi^2}{2\sqrt{\Phi^2+1}}s$.

The area of each triangle is $\frac{1}{2}$ (base) (height):

$$Area_{1 \text{ triangle}} = \frac{1}{2}\left(s\right)\left(\frac{\Phi^2}{2\sqrt{\Phi^2+1}}s\right) = \frac{\Phi^2}{4\sqrt{\Phi^2+1}}s^2.$$

Area of pentagon = 5 times the area of each triangle.

$$Area_{\text{pentagon}} = \frac{5\Phi^2}{4\sqrt{\Phi^2+1}}s^2 = 1.720477401\, s^2. \qquad (8.10)$$

On to *Composition of the Pentagon*, where we further examine the interior of the pentagon and its Phi relationships

The Composition of the Pentagon

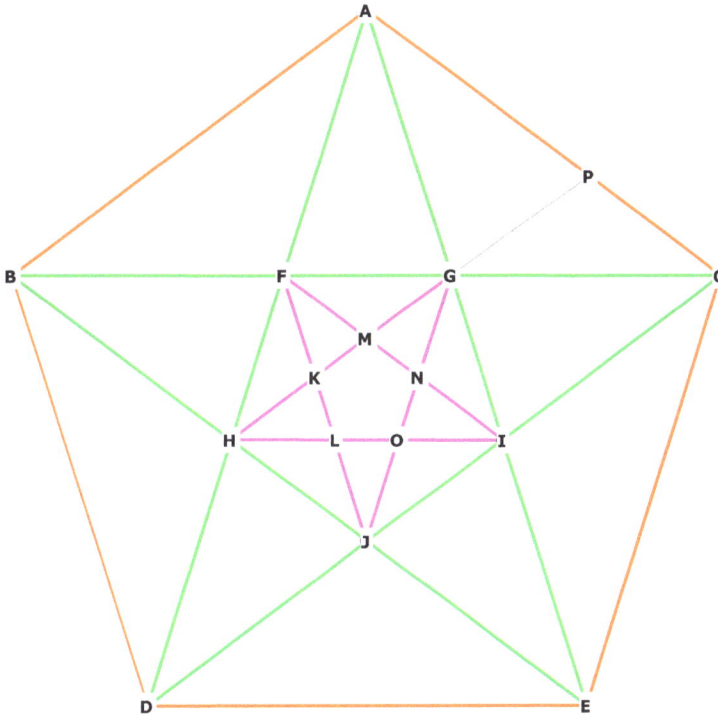

Figure 9.1: Pentagon diagonals

In Figure 9.1, the diagonals of the pentagon (in green) have been marked. These diagonals form another pentagon, GFHJI, which is rotated 36 degrees from the original. By connecting the diagonals of GFHJI, we form another pentagon, KLONM, which is rotated 36 degrees from GFHJI.

Let the sides of the pentagon be of length 1. Then:

$$\overline{BC} = \overline{AD} = \overline{AE} = \overline{CD} = \overline{BE} \quad = \qquad\qquad\qquad\qquad\qquad \Phi$$

$$\overline{AB} = \overline{BC} = \overline{CD} = \overline{DE} = \overline{EA} \quad = \qquad\qquad\qquad\qquad\qquad 1$$

$$\overline{BF} = \overline{GC} = \overline{BH} = \overline{JE} = \overline{AF} \quad = \quad \overline{HD} = \overline{DJ} = \overline{IC} = \overline{AG} = \overline{IE} = \frac{1}{\Phi}$$

$$\overline{FG} = \overline{FH} = \overline{HJ} = \overline{JI} = \overline{GI} \quad = \qquad\qquad\qquad\qquad\qquad \frac{1}{\Phi^2}$$

$$\overline{FK} = \overline{FM} = \overline{GM} = \overline{GN} = \overline{IN} \ = \ \overline{IO} = \overline{JO} = \overline{JL} = \overline{HL} = \overline{HK} = \frac{1}{\Phi^3}$$

$$\overline{OL} = \overline{LK} = \overline{KM} = \overline{MN} = \overline{NO} = \qquad\qquad\qquad\qquad\qquad \frac{1}{\Phi^4}$$

and so on.

Every one of the diagonals of the pentagon divides the other diagonals in Mean and Extreme (Phi) ratio!

Because the sides and the diagonals of the pentagon are in Phi relationship, every one of the triangles within the pentagon is a Phi ratio triangle. The number of these triangles is too numerous to mention, but they are all either 36–36–108 triangles (such as $\triangle AFB$ and $\triangle GIJ$) or 36–72–72 triangles (such as $\triangle DAE$, $\triangle HBF$ and $\triangle GEC$).

This is easily seen if you remember that the exterior angles of the pentagon (such as $\angle BAC$) are all 108°, and the diagonals (such as $\overline{AD}$ and $\overline{AE}$) trisect $\angle BAC$, making $\angle BAF$, $\angle FAG$, and $\angle GAC$ all 36°.

Notice that when the short side of any of these triangles is transferred to a longer side and a point marked, the resulting two triangles are both Phi ratio triangles. For example, observe triangle $\triangle AGC$ in Figure 9.1. $\triangle AGC$ is a 36–36–108 triangle. When the distance $\overline{AG}$ is transferred with the compass to line $\overline{AC}$ at P, two triangles are formed: $\triangle AGP$, a 72–72–36 triangle, and $\triangle PGC$, a 36–36–108 triangle.

If the small pentagon in the center had its diagonals drawn they too would divide themselves in Phi ratio and so the process would continue. Smaller and smaller pentagons would be formed, each one rotating 36 degrees and having sides Φ times smaller than the previous one.

Where does this end? or begin? The answer is, it doesn't!

The pentagon "ratchets" downward into infinite smallness, and similarly ratchets upward, getting larger and larger. This can only occur because division of each line segment is absolutely perfect. Not almost perfect, but totally perfect. There is never any "round-off" error.

Notice also that each large triangle ($\triangle ADE$ for example) is a golden triangle.

The smaller triangles ($\triangle AFG$ for example) are also golden mean triangles.

The pentagon itself can be considered to be a rotating golden mean triangle. If you look at triangle $\triangle ADE$ you can see how it rotates 5 times around the center of the circle (O), each rotation being 72 degrees ($5 \cdot 72 = 360$ for the full circle).

Imagine the pentagon rotating downwards, out of the page, becoming infinitely small. Envision it rotating upwards out of the page, becoming infinitely large. The vertices, as they rotate, form a spiral.

So the growth cycle of the pentagon is like this:

$$\ldots\ldots\frac{1}{\Phi^7},\ \frac{1}{\Phi^6},\ \frac{1}{\Phi^5},\ \frac{1}{\Phi^4},\ \frac{1}{\Phi^3},\ \frac{1}{\Phi^2},\ \frac{1}{\Phi},\ 1,\ \Phi,\ \Phi^2,\ \Phi^3,\ \Phi^4,\ \Phi^5,\ \Phi^6,\ \Phi^7,\ \ldots\ldots,$$

On to *The Icosahedron*

Part III

Now that we understand the pentagon and Phi, we can begin to make sense of the Icosahedron and the Dodecahedron. Although the pentagonal nature of the dodecahedron is obvious, it turns out that the icosahedron is essentially pentagonal as well!

Chapter 10

The Icosahedron

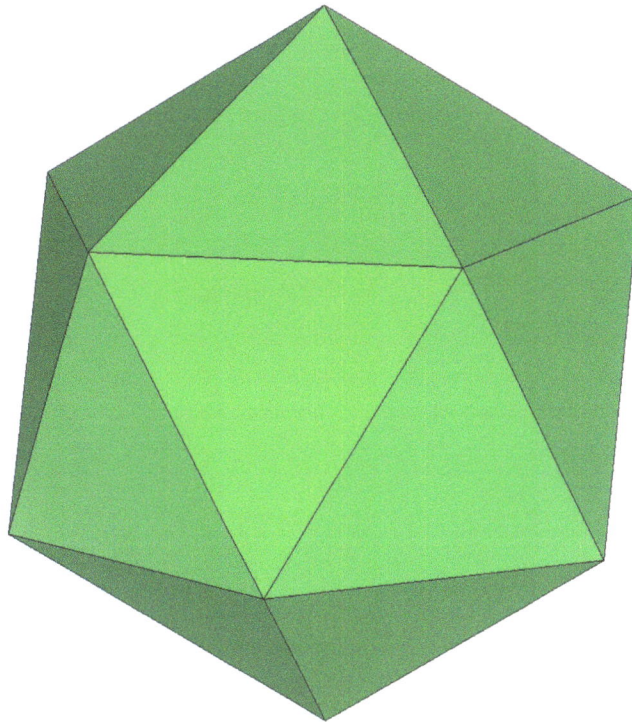

Figure 10.1: Icosahedron

The icosahedron has 12 vertices, 20 faces and 30 sides. It is one of the most interesting and useful of all polyhedra. Buckminster Fuller based his designs of geodesic domes around the icosahedron.

The icosahedron is built around the pentagon and the golden section. At first glance this claim may seem absurd, since every face of the icosahedron is an equilateral triangle. It turns out, however, that the triangular faces of the icosahedron result from its pentagonal nature.

We'll display three views of this polyhedron:

Figure 10.2 shows two of the internal pentagons of the icosahedron, LGHJK, and ABCEF.

Figure 10.3 helps to show that the icosahedron is actually an interlocking series of pentagons. Notice the exterior pentagons at AJKCD, ABCEF, and EFHIL, and LKBDE.

Figure 10.4 shows a two dimensional "shadow" of the icosahedron from the top down. You can see that the outer edges form a perfect decagon, formed by the two pentagons CDFGL and ABKIH.

In fact, every vertex of the icosahedron is the vertex of a pentagon.

OK, lets go through the usual analysis and then get on to the interesting stuff!

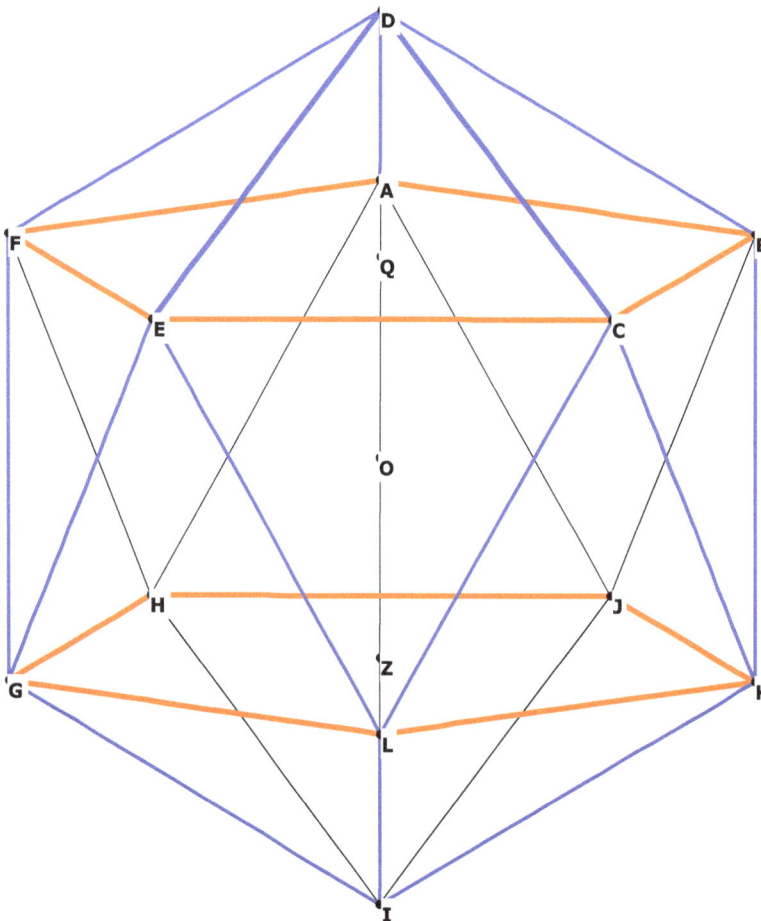

Figure 10.2: Icosahedron, showing internal pentagons

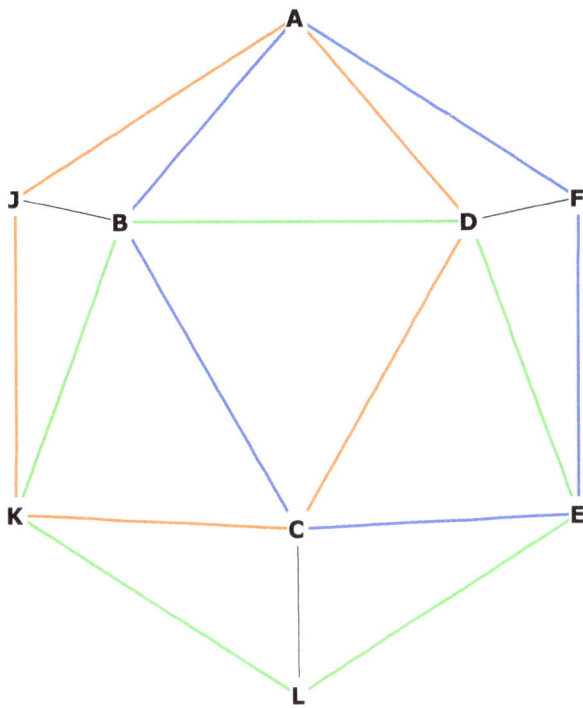

Figure 10.3: Icosahedron showing pentagonal faces

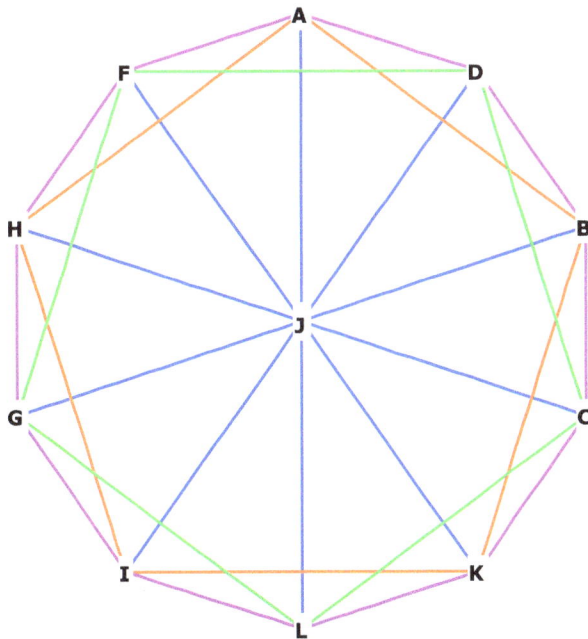

Figure 10.4: Icosahedrom, 2-D shadow

First, lets calculate the volume of the icosahedron.

10.1. Icosahedron Volume

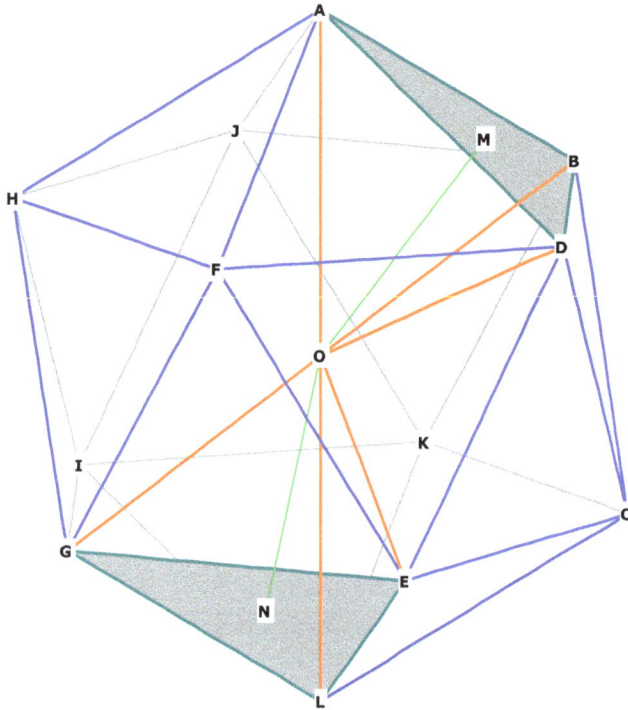

Figure 10.5: Icosahedron showing two pyramids O-ABD and O-ELG

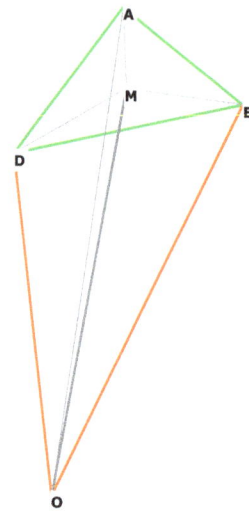

Figure 10.6: An internal pyramid of the icosahedron

Using the pyramid method we have 20 equilateral triangular faces which serve as the base to a pyramid whose topmost point is the centroid of the icosahedron. All 20 faces will be connected to O to form 20 pyramids, thus calculating the volume of the icosahedron.

The volume of each pyramid is $\frac{1}{3}$ (area of base) (pyramid height).

The area of the base is the area of the equilateral triangle $\triangle ADB$.

The height of the pyramid is $\overline{OM}$.

From *The Equilateral Triangle* we know this area is $\frac{\sqrt{3}}{4} s^2$.

All vertices of the icosahedron (as with all five of the regular solids) lie upon the surface of a sphere that encloses it. The radius of the circumsphere is O to any vertex, in this case,

$$r = \overline{OA} = \overline{OB} = \overline{OD} = 1.$$

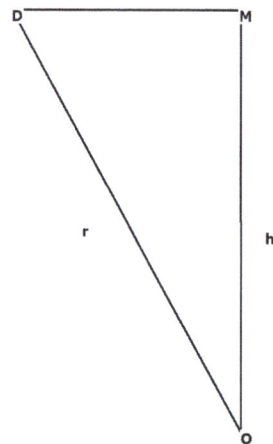

Figure 10.7: Icosahedron pyramid height

The height of the pyramid is $h = \overline{OM}$. In order to find h we need to recognize that any of the triangles $\triangle OMA$, $\triangle OMB$, $\triangle OMD$ are right.

This is because $\overline{OM}$ is perpendicular to the plane of triangle $\triangle ABD$ by construction.

$\overline{AB} = \overline{BD} = \overline{AD} =$ side of icosahedron $= s$.

We know, from *The Equilateral Triangle*, that $\overline{DM} = \frac{1}{\sqrt{3}}s$.

In order to find h, we need to find $\overline{OD} = r$, in terms of the side s of the icosahedron.

To do that, we have to recognize one of the basic geometric properties of the icosahedron.

Take a look at Figure 10.2. $\overline{DI}$, $\overline{BG}$, and $\overline{FK}$ are all diameters of the enclosing sphere around the icosahedron.

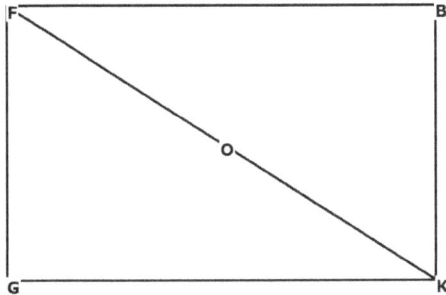

Figure 10.8: Icosahedron, diameter and radius of circumsphere

Notice rectangle BFGK and notice that $\overline{BG}$, $\overline{FK}$ are both diagonals of it. Now notice that both $\overline{FG}$ and $\overline{BK}$ are both diagonals of the two pentagonal planes shown in orange.

We know from *Composition of the Pentagon* that the diagonal of a pentagon is Φ times the side of the pentagon.

Therefore $\overline{FB} = \overline{GK} = \Phi \cdot$ side of icosahedron, since each side of the pentagon is a side of the icosahedron.

$\overline{FB} = \Phi \cdot s$.

$\overline{FK}$ is the diameter of the enclosing sphere around the icosahedron. $\overline{OF} = \overline{OK}$ is the radius r, which we are trying to find.

$\overline{FG} = \overline{BK}$ is the side s of the icosahedron.

$$d^2 = \overline{FK}^2 = \overline{FB}^2 + \overline{BK}^2 = \Phi^2 s^2 + s^2 = (\Phi^2 + 1)s^2.$$

$$d = \overline{FK} = \sqrt{\Phi^2 + 1}\,s,$$

$$r = \frac{d}{2} \text{ and } r = \frac{\sqrt{\Phi^2 + 1}}{2}s. \tag{10.1}$$

Now we can find h, the pyramid height.

Going back to Figure 10.7 and Figure 10.8, we can write:

$$h^2 = \overline{OM}^2 = r^2 - \overline{DM}^2$$

$$= \frac{(\Phi^2 + 1)}{4} - \frac{1}{3} = \frac{3(\Phi^2 + 1) - 4}{12}$$

$$= \frac{3(\Phi^2 - 1)}{12} = \frac{\Phi^4}{12} s^2.$$

$$h = \frac{\Phi^2}{2\sqrt{3}} s \text{ is the height of the icosahedron pyramid.} \tag{10.2}$$

$$Volume_{1pyramid} = \frac{1}{3} \text{ (area of base) (height)} = \frac{1}{3} \left(\frac{\sqrt{3}}{4} s^2 \right) \left(\frac{\Phi^2}{2\sqrt{3}} s \right).$$

$$\text{So } V_{1pyramid} = \frac{\Phi^2}{24} s^3.$$

There are 20 pyramids, 1 for each face so

$$Volume_{icosahedron} = \frac{5\Phi^2}{6} s^3 = 2.181694991 \, s^3. \tag{10.3}$$

10.2. Surface Area of Icosahedron

The surface area of the icosahedron is just the area of 1 face times 20 faces. The area of each face is, from above, $\frac{\sqrt{3}}{4} s^2$.

$$\text{Surface area}_{icosahedron} = 20 \left(\frac{\sqrt{3}}{4} s^2 \right) = 5\sqrt{3} \, s^2 = 8.660254038 \, s^2. \tag{10.4}$$

10.3. Radius to Side

We have already noted the relationship between the radius of the enclosing sphere and the side of the icosahedron:

$$r = \frac{\sqrt{\Phi^2 + 1}}{2} s, s = \frac{2}{\sqrt{\Phi^2 + 1}} r.$$

$$r = .951056517 \, s, \ s = 1.051462224 \, r. \tag{10.5}$$

The side or edge of the icosahedron is slightly larger than the radius.

10.4. Central Angle of Icosahedron

The central angle of the icosahedron, $\angle DOB$, can be seen clearly from Figure 10.7, and we diagram it in Figure 10.9.

$\overline{OD} = \overline{OB} = r = \frac{\sqrt{\Phi^2+1}}{2} s$. $\overline{DB}$ is the side of the icosahedron, s.

$$\sin(\angle XOB) = \frac{\overline{XB}}{\overline{OB}} = \frac{\frac{1}{2}}{\frac{\sqrt{\Phi^2+1}}{2}} = \frac{1}{\sqrt{\Phi^2+1}}.$$

$$\angle XOB = \arcsin\left(\frac{1}{\sqrt{\Phi^2+1}}\right) = 31.7174744°.$$

$$\angle DOB = 2\left(\angle XOB\right).$$

We recognize triangle $\triangle OXB$ as our old friend the Phi Right Triangle. From this we know that $\frac{\overline{OX}}{\overline{XB}} = \Phi$.

$$\text{Central angle} = 2\left(\angle XOB\right) = 63.4349488°.$$

$$\text{Surface angles} = 60° \qquad\qquad (10.6)$$

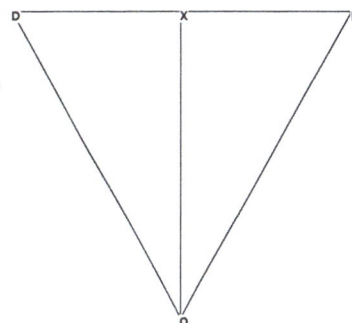

Figure 10.9: Central angle of icosahedron

10.5. Dihedral Angle of Icosahedron

The dihedral angle of the icosahedron is the angle formed by the intersection of two planes:

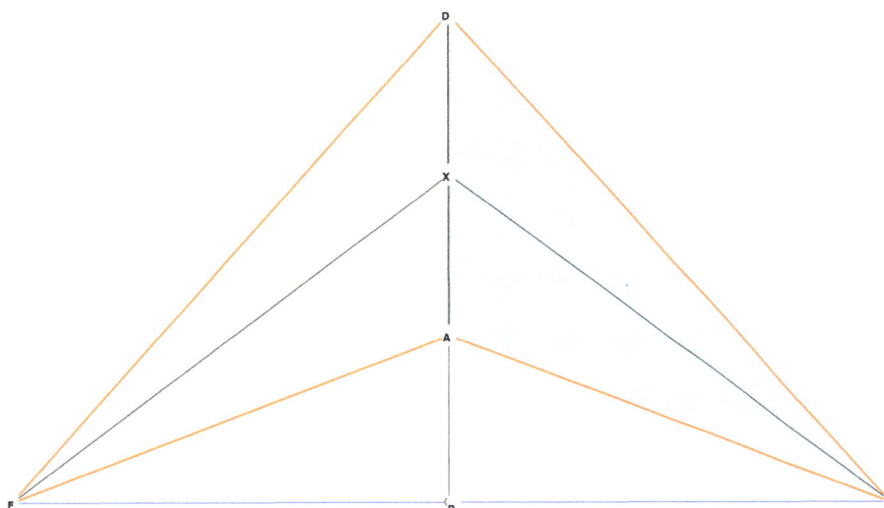

Figure 10.10: Dihedral angle of icosahedron

The intersection of the two faces DFA and ABD forms the dihedral angle $\angle FXB$. (See Figure 10.6 as well).

We know, from *The Equilateral Triangle*, that $\overline{FX} = \overline{XB} = \frac{\sqrt{3}}{2}s$.

$\overline{FB}$ = a diagonal of one of the pentagons. This can be seen in Figure 10.5 as the diagonal of pentagon ABCEF.

Therefore $\overline{FB} = \Phi s$ and $\overline{PB} = \frac{\Phi}{2} s$.

Triangles $\triangle XPB$ and $\triangle XPF$ are right by construction, so

$\angle PXB = \frac{1}{2}\angle FXB$.

$$\sin(\angle PXB) = \frac{\overline{PB}}{\overline{XB}} = \frac{\frac{\Phi}{2}}{\frac{\sqrt{3}}{2}} = \frac{\Phi}{\sqrt{3}}.$$

$$\angle PXB = \arcsin\left(\frac{\Phi}{\sqrt{3}}\right) = 69.09484258°$$

$$\text{Dihedral angle} = 2\,(\angle PXB) = 138.1896852° \tag{10.7}$$

10.6. Centroid Distances

Now let's figure out the distances from the centroid of the icosahedron to any vertex, to any mid-face, and to any mid-side.

We already have the first two (see Figure 10.11). $\overline{OD} = \overline{OA} = \overline{OB} = r = \frac{\sqrt{\Phi^2+1}}{2}s.$

$$\overline{OM} = h = \frac{\Phi^2}{2\sqrt{3}}s.$$

Now we need to find, for example, $\overline{OX}$.

If we lay a 3D model of the icosahedron on one of its sides, we can see that a line through the centroid O is perpendicular to that side. So the triangle $\triangle OXB$ is right.

We know $\overline{AB} = s$, so $\overline{XB} = \frac{1}{2}s$. $\overline{OB} = r$, so

$$\overline{OX}^2 = \overline{OB}^2 - \overline{BX}^2$$

$$= \frac{\Phi^2+1}{4}s^2 - \frac{1}{4}s^2$$

$$= \frac{\Phi^2+1-1}{4}s^2 = \frac{\Phi^2}{4}s^2.$$

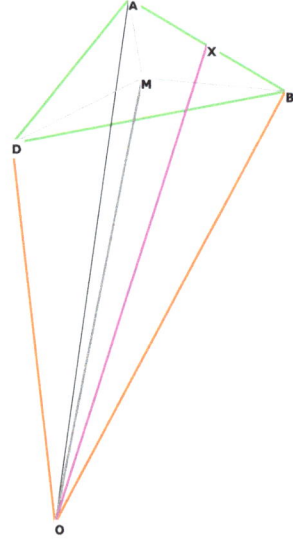

Figure 10.11: Distances from centroid of icosahedron.

$$\overline{OX} = \frac{\Phi}{2}s. \tag{10.8}$$

$$\text{distance from centroid to mid-face} = \frac{\Phi^2}{2\sqrt{3}}s = 0.755761314\,s. \tag{10.9}$$

distance from centroid to mid-side $= \dfrac{\Phi}{2} s = 0.809016995\, s.$ (10.10)

distance from centroid to a vertex $= \dfrac{\sqrt{\Phi^2 + 1}}{2} s = 0.951056517\, s.$ (10.11)

Now lets get to the interesting stuff!

10.7. Analysis of the Pentagonal Cap

Going back to Figure 10.2, we can see that the icosahedron is composed of interlocking pentagonal "caps."

Look at D-ABCEF and I-GHJKL to see this more clearly. Of course, EVERY vertex of the icosahedron is the top of a pentagonal cap, not just D and I.

Let's analyze this cap:

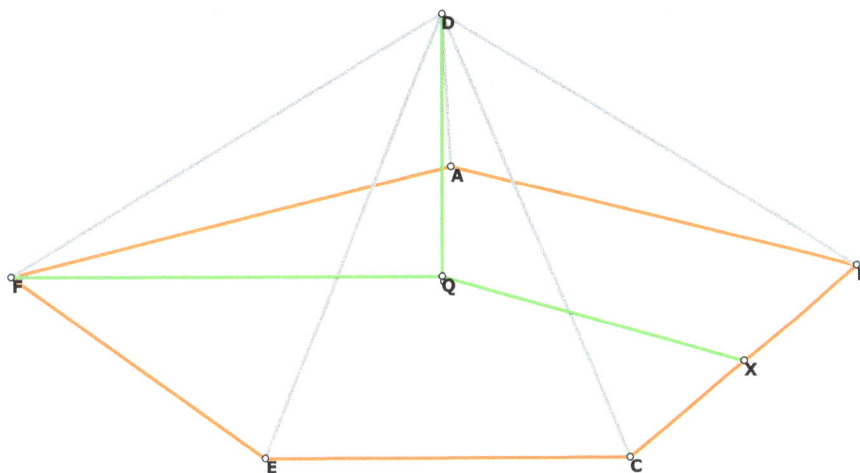

Figure 10.12: Pentagonal cap of icosahedron

Triangle $\triangle DQF$ is right, by construction. D is directly over the point Q in the icosahedron.

How far off the plane of pentagon AFECB is D?

In other words, what is the distance $\overline{DQ}$?

First, realize there is a circle around pentagon AFECB, even though I haven't draw it here. $\overline{QF}$ is the radius of that circle. We know from *Construction of the Pentagon, Part 2* that this radius

$r = \overline{FQ} = \frac{\Phi}{\sqrt{\Phi^2+1}}\, s.$ Here, s is the side of the icosahedron.

$\overline{DF}$ is a side of the icosahedron, so $\overline{DF} = s$. Therefore we write

$$\overline{DQ}^2 = \overline{DF}^2 - \overline{FQ}^2$$

$$= s^2 - \frac{\Phi^2}{\Phi^2 + 1}s^2$$

$$= \frac{(\Phi^2 + 1) - \Phi^2}{\Phi^2 + 1}s^2 = \frac{1}{\Phi^2 + 1}s^2.$$

$$\overline{DQ} = \frac{1}{\sqrt{\Phi^2 + 1}}s. \tag{10.12}$$

Hmmm, this is looking interesting. Lets compare $\overline{FQ}$ to $\overline{DQ}$.

$$\frac{\overline{FQ}}{\overline{DQ}} = \frac{\frac{\Phi}{\sqrt{\Phi^2+1}}}{\frac{1}{\sqrt{\Phi^2+1}}} = \Phi! \tag{10.13}$$

In order to form equilateral triangles with D and the other vertices of the pentagon, D has to be raised off the pentagonal plane AFECB by the division of the radius of the pentagon $(\overline{FQ})$ in Mean and Extreme Ratio.

The triangle $\triangle DQF$ is therefore a Phi based triangle, specifically, a 1, Φ, $\sqrt{\Phi^2 + 1}$ triangle. From *The Phi Right Triangle* we know that $\triangle DFQ = 31.71747441°$.

Is this surprising? It was to me! I didn't expect something that was composed entirely of equilateral triangles to have any relationship to Φ. An equilateral triangle is $\sqrt{3}$ geometry, Φ is $\sqrt{5}$ geometry.

10.8. Distances to Pentagonal Planes

In Figure 10.13 we have Figure 10.2 basically, with the mid-face points of the two internal pentagons marked off as Q and Z.

We have already seen that $\overline{DQ}$ is $\frac{1}{\sqrt{\Phi^2+1}}$ with respect to the side of the icosahedron.

That means $\overline{IZ} = \frac{1}{\sqrt{\Phi^2+1}} s$ as well, because $\overline{IZ} = \overline{DQ}$.

What about $\overline{OQ} = \overline{OZ}$? and $\overline{QZ}$? How do all of these distances relate to the diameter of the enclosing sphere, $\overline{DI}$?

Remember that the distance $\overline{DI} = \overline{FK} = \overline{BG}$, etc., is the diagonal of any of the Φ rectangles of which the icosahedron is composed. One of these Φ rectangles is clearly visible in Figure 10.13 as BFGK. We know this is a Φ rectangle because it is the diagonal of the pentagon ABCEF, which sides are the sides of the icosahedron.

In fact, if you place three of these rectangles perpendicular to each other, the 12 corners of the three rectangles are the vertices of the icosahedron!

$\overline{FB} = \Phi$.

I have copied Figure 10.14 from Figure 10.8 above. $\overline{FK}$ = diameter, $\overline{OF}$ = radius of enclosing sphere.

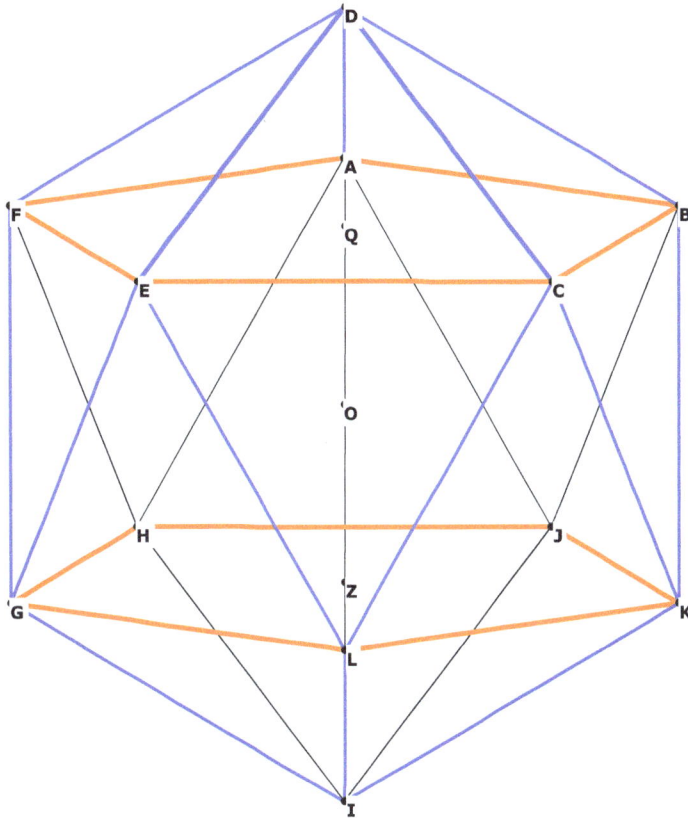

Figure 10.13: Icosahedron with mid-face points of pentagonal planes

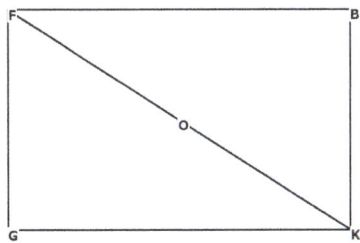

Figure 10.14: Phi rectangle BFGK

We know from above that

$$d = \overline{FK} = \overline{DI} = \sqrt{\Phi^2 + 1}\, s.$$

We have already figured out $\overline{DQ} = \overline{ZI}$. So $\overline{QZ} = \overline{DI} - 2\left(\overline{DQ}\right)$.

$$\overline{QZ} = \sqrt{\Phi^2 + 1} - \left(2\left[\frac{1}{\sqrt{\Phi^2 + 1}}\right]\right)$$

$$= \frac{\Phi^2 + 1 - 2}{\sqrt{\Phi^2 + 1}} = \frac{(\Phi^2 - 1)}{\sqrt{\Phi^2 + 1}} = \frac{\Phi}{\sqrt{\Phi^2 + 1}}\, s.$$

Now,

$$\frac{\overline{QZ}}{\overline{DQ}} = \frac{\frac{\Phi}{\sqrt{\Phi^2+1}}}{\frac{1}{\sqrt{\Phi^2+1}}} = \Phi! \tag{10.14}$$

$\overline{DZ}$ is divided in EMR at Q (See Figure 10.13).

One interesting fact appears here: $\overline{FQ} = \overline{QZ}$. This means that the distance from one pentagonal plane to the other is precisely the radius of the circle that encloses the pentagon ABCEF.

10.8.1. WHAT IS $\overline{OQ}$, THE DISTANCE BETWEEN THE CENTROID AND THE TWO PENTAGONAL PLANES?

It looks like $\overline{OQ}$ is just one-half that of $\overline{QZ}$, or $\frac{\Phi}{2\sqrt{\Phi^2+1}}\, s$. But is it? Let's find out.

We know $\overline{OD} = r = \frac{\sqrt{\Phi^2+1}}{2}$ from above.

$$\overline{OQ} = \overline{OD} - \overline{DQ} = \frac{\sqrt{\Phi^2+1}}{2} - \frac{1}{\sqrt{\Phi^2+1}} = \frac{\Phi^2 + 1 - 2}{2\sqrt{\Phi^2+1}} = \frac{(\Phi^2-1)}{2\sqrt{\Phi^2+1}} = \frac{\Phi}{2\sqrt{\Phi^2+1}}\, s. \tag{10.15}$$

$\overline{OQ} = \frac{\Phi}{2\sqrt{\Phi^2+1}}\, s.$ We could have gotten this more easily from the fact that $\overline{DQ} = \frac{1}{\sqrt{\Phi^2+1}}s.$

$\overline{OQ}$ is exactly one-half $\overline{QZ}$.

Also,

$$\frac{\overline{OQ}}{\overline{DQ}} = \frac{\Phi}{2}\, s = 0.951056517. \tag{10.16}$$

10.9. Table of Relationships

(Let $\overline{DQ} = 1$)

See Table 10.1 below. $\overline{DZ}$ is divided in EMR by Q, $\overline{IQ}$ is divided in EMR by Z.

All of these relationships come from the pentagon!

Table 10.1: Central axis (diameter) distances. Diameter of enclosing sphere = $\overline{DI}$ graph reads vertically, by column

D		D	
1			
Q	Q		Q
$\Phi/2$		$\Phi + 1 = \Phi^2$	
O	Φ		
$\Phi/2$			$\Phi + 1 = \Phi^2$
Z	Z	Z	
1			
I			I

On the outside of the icosahedron, we see equilateral triangles. But the guts of this polyhedron comes from pentagonal relationships. The equilateral triangles come about from the lifting of the pentagonal "cap" off the pentagonal plane.

There is now no question that the basis for the construction of the icosahedron is the pentagon.

Or is there?

Take a look at this view of the icosahedron:

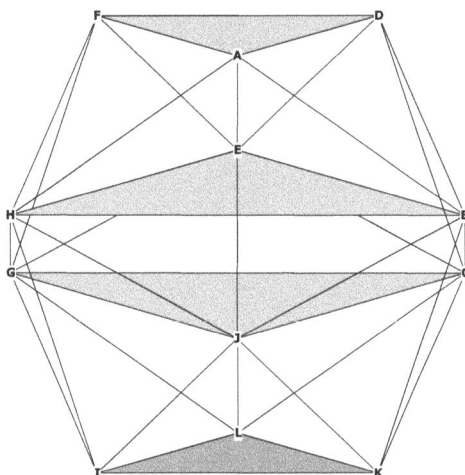

Figure 10.15: Showing the internal triangular planes of the icosahedron

As you can see, the two pentagonal planes in the middle have magically disappeared and become equilateral triangles. All we have done is placed the icosahedron on one of its faces.

It seems that all of our work is wrong, except:

The sides of the equilateral triangles $\triangle EHB$ and $\triangle JGC$ are all diagonals of pentagons!

$\overline{EH}$ is a diagonal of pentagon FEJIH, $\overline{EB}$ is a diagonal of pentagon DEJKB, and $\overline{HB}$ is a diagonal of pentagon AHIKB.

The equilateral triangle faces of the icosahedron are a by-product of how the pentagons interlock. This is shown in Figure 10.16 below:

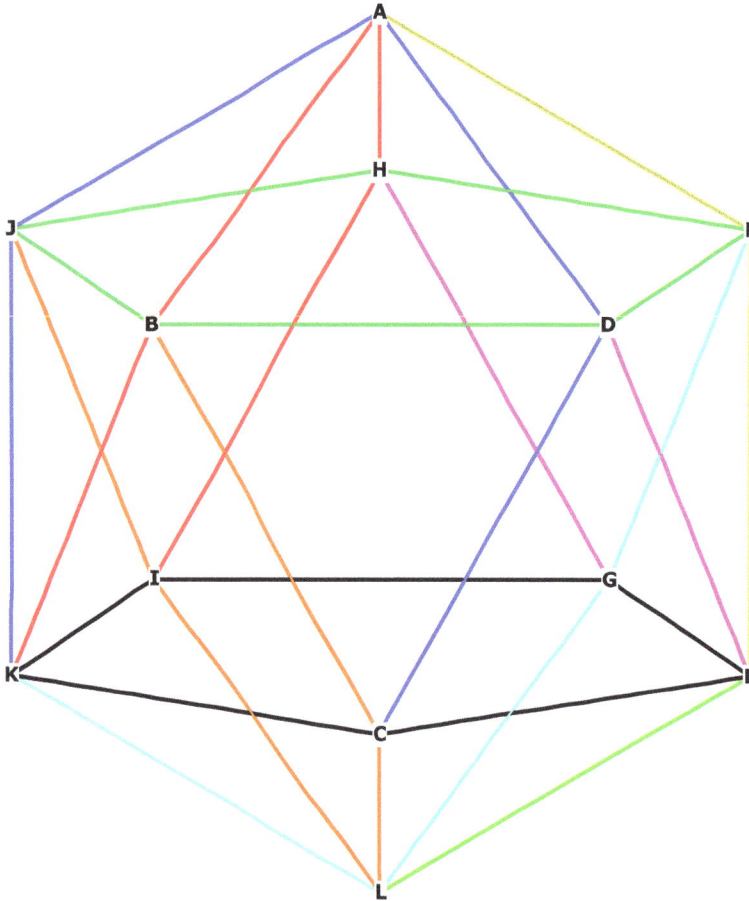

Figure 10.16: Attempting to show the individual pentagons of the icosahedron

Because the pentagons are interlocking, there is duplication of colors, which makes the individual pentagons a bit difficult to identify.

Begin with the two pentagons BDFHJ, in green, and CEGIK, in black.

Insert pentagon AJKCD, 4 of which sides are in blue.

Insert pentagon ABKIH, 4 of which sides are marked in red.

Insert pentagon BCLIJ, 4 of which sides are marked in orange.

Insert pentagon ADEGH, 2 of which sides are marked in magenta.

Insert pentagon AFECB, 2 of which sides are marked in gold.

Insert pentagon CDFGL, 2 of which sides are marked in cyan (sky blue)

The side $\overline{KL}$ is part of the pentagon KLGHJ.

The side $\overline{LE}$ is part of the pentagon ELIHF.

10.10. Icosahedron Reference Charts

Table 10.2: Volume and Surface Area

Volume in terms of s	Volume in Unit Sphere	Surface Area in terms of s	Surface Area in Unit Sphere
$2.181694991 s^3$	$2.53615071 r^3$	$8.660254038 s^2$	$9.574541379 r^2$

Table 10.3: Angles

Central Angle:	Dihedral Angle:	Surface Angle:
$63.4349488°$	$138.1896852°$	$60°$

Table 10.4: Centroid Distances

Centroid To Vertex:	Centroid To Mid-edge:	Centroid To Mid-face:
$1.0 r$	$0.850650809 r$	$0.794654473 r$
$0.951056517 s$	$0.809016995 s$	$0.755761314 s$

Table 10.5: Side to Radius

Side / radius
1.051462224

Chapter 11

The Dodecahedron

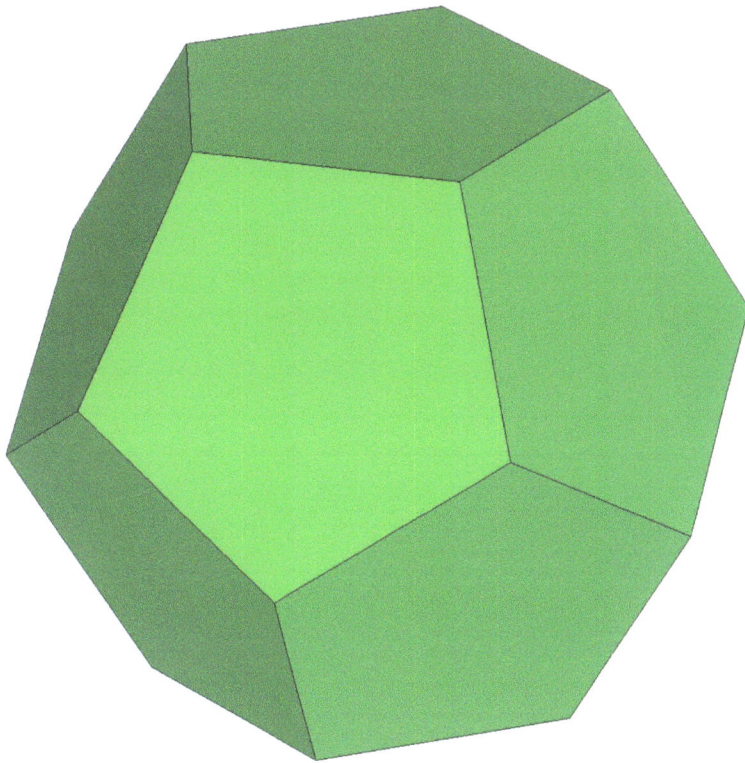

Figure 11.1: Dodecahedron

The dodecahedron has 30 edges, 20 vertices and 12 faces. Dodeca is a prefix meaning "twelve."

11.1. Dodecahedron Overview

The dodecahedron is composed entirely of pentagons.

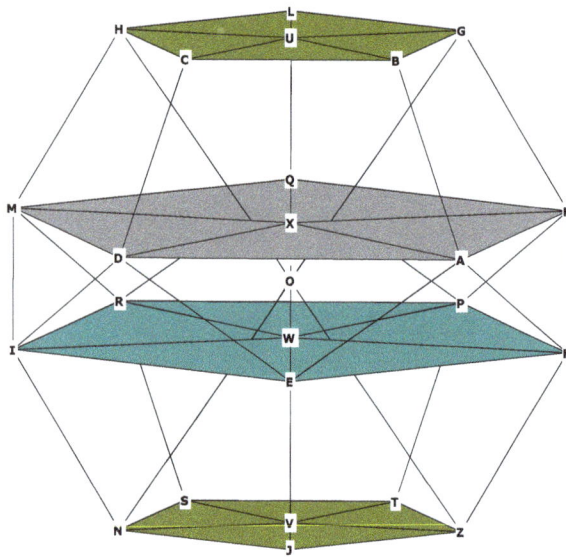

Figure 11.2: Dodecahedron showing internal pentagons

The dodecahedron is pentagonal both inside and out, as can be seen from Figure 11.2. Like the icosahedron, it has many golden section relationships, which we shall see.

The dodecahedron is even more versatile than the icosahedron. The icosahedron contains $\sqrt{3}$ and $\sqrt{5}$ geometry, but the dodecahedron contains $\sqrt{2}$, $\sqrt{3}$ and $\sqrt{5}$ geometry!

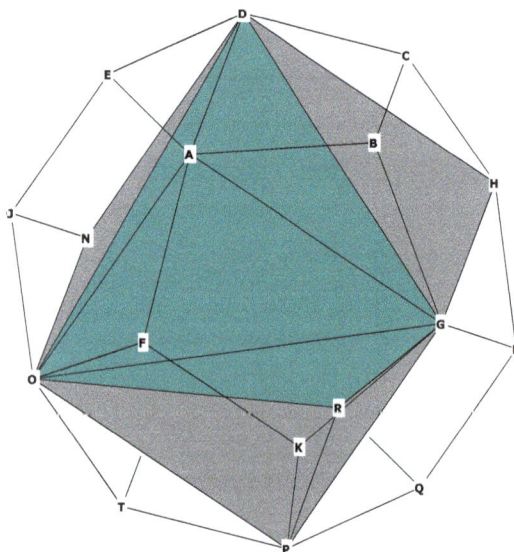

Figure 11.3: Cube and tetrahedron in dodecahedron. Cube in gray, tetrahedron in green

The view of the dodecahedron in Figure 11.3 is significant in that it shows the 2-dimensional shadow of the decagon. The decagon itself is based upon the pentagon, the building block of the dodecahedron.

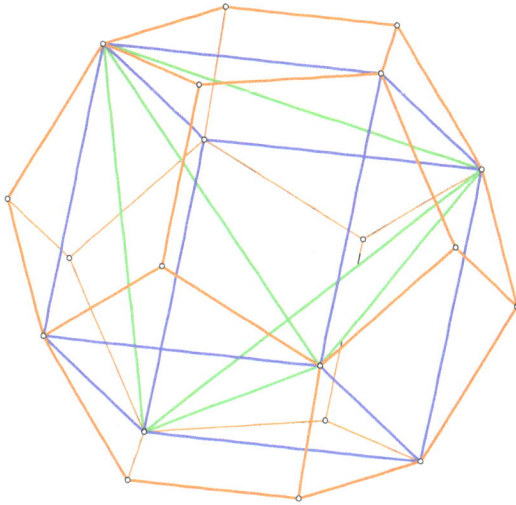

Figure 11.4: Another view of tetrahedron (green) inside cube (blue) inside dodecahedron (orange)

Figures 11.3 and 11.4 show how a cube and a tetrahedron can be placed inside a dodecahedron, directly upon its vertices.

The cube, octahedron and tetrahedron are all based on root 2 and root 3 geometry The relationship of the side of the cube to the radius of its enclosing sphere is $r = \frac{\sqrt{3}}{2}s$. For the tetrahedron, $r = \frac{2\sqrt{2}}{\sqrt{3}}s$. For the octahedron, $r = \frac{1}{\sqrt{2}}s$.

The dodecahedron is capable of elegantly sustaining these $\sqrt{2}$ and $\sqrt{3}$ relationships, along with its own many $\sqrt{5}$ relationships.

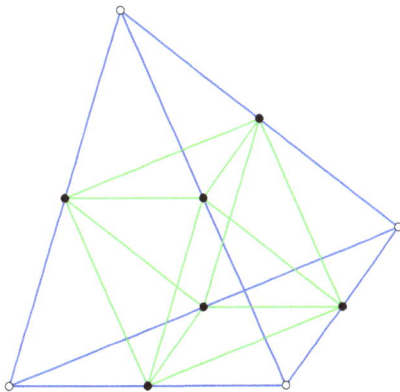

Figure 11.5: Octahedron inside tetrahedron

Notice that the octahedron fits precisely on the bisected sides of the tetrahedron.

The icosahedron cannot contain any of the other five solids "nicely" on its vertices.

The icosahedron and the dodecahedron are duals (as are the cube and the octahedron). By 'dual' is meant that if you put a vertex in the middle of every face and connect the lines, you get the dual.

By placing a vertex at the middle of each face of the dodecahedron you get an icosahedron, and vice-versa. Figure 11.6 shows the dual nature of the icosahedron and dodecahedron.

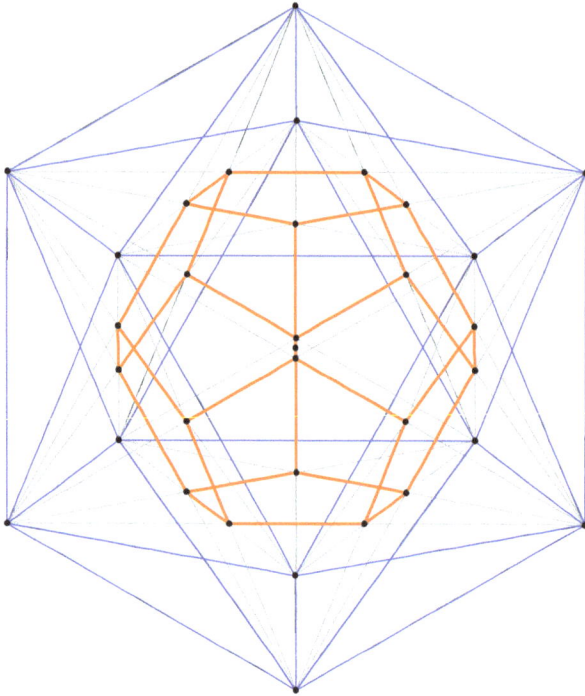

Figure 11.6: Duals — dodecahedron inside icosahedron

To create the dodecahedron, all we did was draw lines from each vertex of the icosahedron to every other vertex. The vertices of the dodecahedron are at the intersection points. We could just as easily have found the vertices of the dodecahedron by drawing lines on every triangular face of the icosahedron. Where those lines intersect is the center of the face, and a vertex of the dodecahedron. That occurs because the dodecahedron has 12 faces and the icosahedron has 12 vertices. I've shown the dual this way because the diagram is less cluttered.

Now for the standard analysis.

11.2. Volume of Dodecahedron

We will use the pyramid method.

There are 12 pentagonal pyramids, 1 for each face, each pyramid beginning at O, the centroid. See Figure 11.2 and Figure 11.7.

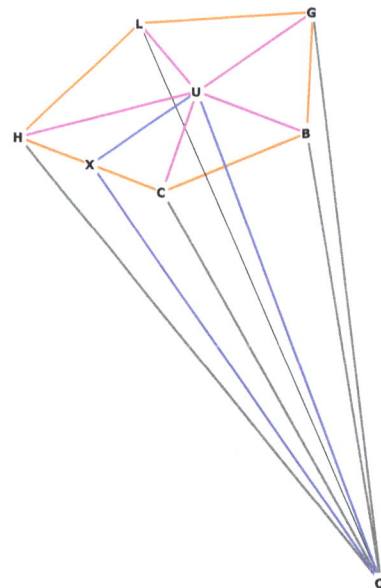

Figure 11.7: One pyramid on face BCHLG

The volume of any n-sided pyramid is $\frac{1}{3}$ (area of base) (pyramid height).

First we need to get the area of the base, which is the area of each pentagonal face:

The area of the pentagon is the area of the 5 triangles which compose it.

From section 8.8 *Area of the Pentagon* we know

$$a = \frac{5\Phi^2}{4\sqrt{\Phi^2 + 1}}s^2 = 1.720477401\,s^2.$$

Now we need to find the height of the pyramid, $\overline{OU}$. To do that, we need to find the distance from O to a vertex, lets say, $\overline{OH}$. This distance will be the hypotenuse of the right triangle $\triangle OUH$. Since we already know $\overline{UH}$, we can then get $\overline{OU}$ from the good ol' Pythagorean Theorem.

Imagine a sphere surrounding the dodecahedron and touching all of its vertices (called the circumsphere). $\overline{OH}$ is just the radius of the circumsphere. If you look at Figure 11.8, $\overline{HOZ} = \overline{GON} = \overline{UOV}$ = diameter.

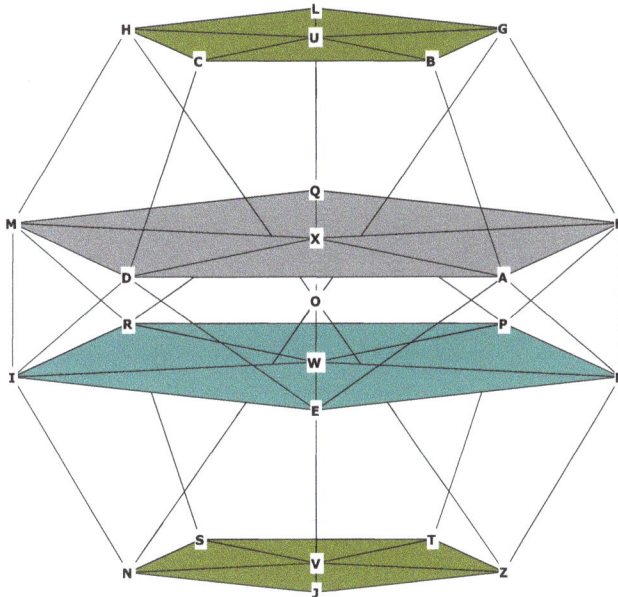

Figure 11.8: Dodecahedral planes

Now look at the rectangle MIFK. The diagonal of it, $\overline{MF}$, is also a diameter ($\overline{MF} = \overline{HZ}$). Notice that the long sides of the rectangle, $\overline{MK}$, and $\overline{IF}$, are diagonals of the two large pentagons ADMQK and FEIRP.

We know from Chapter 9 *The Composition of the Pentagon* that the diagonal of a pentagon is $\Phi \cdot$ side of pentagon.

We also can see from Figure 11.8 and that the sides of the large pentagons themselves are diagonals of the pentagonal faces of the dodecahedron! (For instance, $\overline{DA}$ is a diagonal of the face ABCDE). That means each side of the large pentagons is $\Phi \cdot s$ and that $\overline{MK}$ (or any of the diagonals of a large pentagon) is $\Phi \cdot \Phi \cdot s$.

So $\overline{MK} = \Phi^2 s$.

In fact, like the icosahedron, the dodecahedron is composed of rectangles divided in Extreme and Mean Ratio. In the icosahedron, we found these rectangles to be Φ rectangles.

In the dodecahedron, they are Φ^2 rectangles.

In rectangle MKIF, $\overline{MK} = \overline{IF} = \Phi^2 s, \overline{MI} = \overline{KF} = s$, as shown in Figure 11.9.

M Φ^2 s   K

s

I   F

Figure 11.9: Showing vertices of the Φ^2 rectangle MKIF.

There are 30 sides to the dodec, and therefore 15 different Φ^2 rectangles.

All of this as explanation of finding the distance $\overline{OH}$ from Figure 11.7! Because we are not using trigonometry, we need $\overline{OH}$ in order to get the pyramid height, $\overline{OU}$ in Figure 11.7. Notice that in Figure 11.8, MHFZ is also a Φ^2 rectangle and that $\overline{HZ}$ is the diagonal of it. If we can find $\overline{HZ}$, then $\overline{OH}$ is just one-half of that.

$$d^2 = \overline{HZ}^2 = \overline{MZ}^2 + \overline{HM}^2 = \Phi^4 + 1 = (3\Phi + 2) + 1 = 3\Phi + 3 = 3(\Phi + 1) = 3\Phi^2.$$

$$\text{diameter of circumsphere} = \overline{HZ} = \sqrt{3}\Phi\, s. \qquad (11.1)$$

$$\overline{OH} = r = \frac{\sqrt{3}\Phi}{2}\, s, \ s = \frac{2}{\sqrt{3}\Phi} r.$$

Now we can find $\overline{OU}$, the height of the pyramid.

From section 8.8 *Area of the Pentagon* we know the distance mid-face to any vertex of a pentagon $= \frac{\Phi}{\sqrt{\Phi^2+1}}\, s$.

So $\overline{UH} = \frac{\Phi}{\sqrt{\Phi^2+1}}$.

$$
\begin{aligned}
h^2 = \overline{OU}^2 &= \overline{OH}^2 - \overline{UH}^2 \\
&= \frac{3\Phi^2}{4} s^2 - \frac{\Phi^2}{\Phi^2 + 1} \\
&= \frac{3\Phi^4 + 3\Phi^2 - 4\Phi^2}{4(\Phi^2 + 1)} \\
&= \frac{3\Phi^4 - \Phi^2}{4(\Phi^2 + 1)} \\
&= \frac{\Phi^2(3\Phi^2 - 1)}{4(\Phi^2 + 1)} \\
&= \frac{\left(\Phi^4\right)\left(\Phi^2\right)}{4(\Phi^2 + 1)} \\
&= \frac{\Phi^6}{4(\Phi^2 + 1)}.
\end{aligned}
$$

$$h = \overline{OU} = \frac{\Phi^3}{2\sqrt{\Phi^2 + 1}} s. \tag{11.2}$$

The volume of 1 pyramid $= \dfrac{1}{3}$ (area of base) (pyramid height)

$$= \frac{1}{3} \left(\frac{5\Phi^2}{4\sqrt{\Phi^2 + 1}} s^2 \right) \left(\frac{\Phi^3}{2\sqrt{\Phi^2 + 1}} s \right)$$

$$= \frac{5\Phi^5}{24(\Phi^2 + 1)} s^3.$$

$$Volume_{dodeca} = 12 \cdot Volume_{1\ pyramid} = \frac{5\Phi^5}{2(\Phi^2 + 1)} s^3 = 7.663118963\ s^3. \tag{11.3}$$

Or

$$Volume_{dodeca} = \frac{5\Phi^5}{2(\Phi^2 + 1)} s^3 = \frac{5\Phi^5}{2(\Phi^2 + 1)} \left(\frac{2}{\sqrt{3\Phi}} r \right)^3$$

$$= \frac{20\Phi^2}{3\sqrt{3}(\Phi^2 + 1)} = 2.78516386\ r^3. \tag{11.4}$$

Note that (from Figure 11.7) $\frac{\overline{OU}}{\overline{UX}} = \Phi$.

11.3. Surface Area of Dodecahedron

The surface area of the dodecahedron is 12 faces times the area of each face =

$$12\, s \left(\frac{5\Phi^2}{4\sqrt{\Phi^2 + 1}} s \right) = \frac{15\Phi^2}{\sqrt{\Phi^2 + 1}} s^2 = 20.64572881\ s^2 \tag{11.5}$$

Or

$$\frac{15\Phi^2}{\sqrt{\Phi^2 + 1}} \left(\frac{2}{\sqrt{3\Phi}} r \right)^2 = \frac{20}{\sqrt{(\Phi^2 + 1)}} r^2 = 10.51462224, r^2 \tag{11.6}$$

11.4. Central Angle of Dodecahedron

From Figure 11.7, the central angle of the dodecahedron is (for example) $\angle HOC$, shown in Figure 11.10:

$\overline{HC}$ = side of dodecahedron, so $\overline{XH}$ = one-half s.

$\overline{OH}$ = radius = one half $\overline{HZ} = \frac{\sqrt{3}\Phi}{2} s$.

$$\sin(\angle XOH) = \frac{\overline{XH}}{\overline{OH}} = \frac{\frac{1}{2}}{\frac{\sqrt{3\Phi}}{2}} = \frac{1}{\sqrt{3\Phi}}.$$

$$\overline{XOH} = \arcsin\left(\frac{1}{\sqrt{3\Phi}} \right) = 20.90515744°.$$

$$\angle HOC = \text{central angle} = 41.81031488°. \tag{11.7}$$

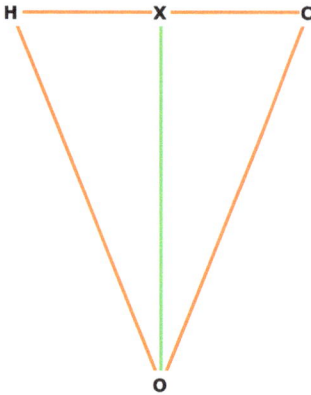

Figure 11.10: Dodecahedron central angle

Because each face of the dodecahedron is a pentagon,

$$\text{the surface angle} = 108°. \tag{11.8}$$

11.5. Centroid Distances

We already have distance from centroid to a vertex, $\overline{OH}$, and the distance from centroid to mid-face, from the previous section (see Figure 11.7).

$$\text{centroid to vertex} = \overline{OH} = r = \frac{\sqrt{3}\Phi}{2} s.$$

$$\text{centroid to mid-face} = \overline{OU} = h = \frac{\Phi^3}{2\sqrt{(\Phi^2 + 1)}} s.$$

Now let's get $\overline{OX}$, the distance from the centroid to any mid- edge.

$$\overline{OX}^2 = \overline{OH}^2 - \overline{HX}^2$$

$$= \frac{3\Phi^2}{4} s^2 - \frac{1}{4} s^2 = \frac{3\Phi^2 - 1}{4} s^2 = \frac{\Phi^4}{4} s^2.$$

$$\overline{OX} = \frac{\Phi^2}{2} s = 1.309016995\, s. \tag{11.9}$$

Let's compare distances:

$$\text{Distance from centroid to mid-face}(h) = \frac{\Phi^3}{2\sqrt{\Phi^2 + 1}} s = 1.113516365\, s. \tag{11.10}$$

$$\text{Distance from centroid to mid-edge} = \frac{\Phi^2}{2} s = 1.309016995\, s. \tag{11.11}$$

$$\text{Distance from centroid to vertex} = \frac{\sqrt{3}\Phi}{2} s = 1.401258539\, s. \tag{11.12}$$

11.6. Dihedral Angle of the Dodecahedron

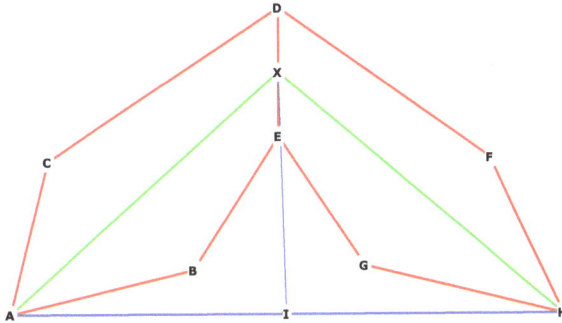

Figure 11.11: Dihedral angle of dodecahedron

The dihedral angle is $\angle AXH$. $\overline{AH}$ is one of the long sides of any of the 15 Φ^2 rectangles which compose the dodec. $\overline{AX}$ and $\overline{HX}$ are the height h of the pentagon.

We know from Chapter 8 *Construction of the Pentagon Part II* that the height h of the pentagon is: $\frac{\Phi\sqrt{\Phi^2+1}}{2}s$.

We know from Figure 11.9 that $\overline{AH}$ is $\Phi^2 s$, so $\overline{IH} = \frac{\Phi^2}{2}s$.

$$\sin(\angle IXH) = \frac{\overline{IH}}{\overline{XH}} = \frac{\frac{\Phi^2}{2}}{\frac{\Phi\sqrt{\Phi^2+1}}{2}} = \frac{\Phi}{\sqrt{\Phi^2+1}}.$$

We recognize this ratio as our friend the Phi Right Triangle with sides in ratio of 1, Φ, $\sqrt{\Phi^2+1}$.

$$\angle IHX = \arcsin\left(\frac{\Phi}{\sqrt{\Phi^2+1}}\right) = 58.28252558°.$$

$$\text{dihedral angle} \angle AXH = 2\left(\angle IXH\right),$$

So

$$\text{dihedral angle} \angle AXH = 116.5650512°. \tag{11.13}$$

11.7. Dodecahedron Planar Distances

Go back to Figure 11.8, shown below as Figure 11.12. We have colored the four internal pentagonal planes of the dodecahedron. U, X, W, and V are the centers of these four planes which line up with the centroid O.

What is the distance $\overline{UX} = \overline{WV}$? What is $\overline{XW}$?

If we can find these out we can figure out more deeply how the dodecahedron is constructed.

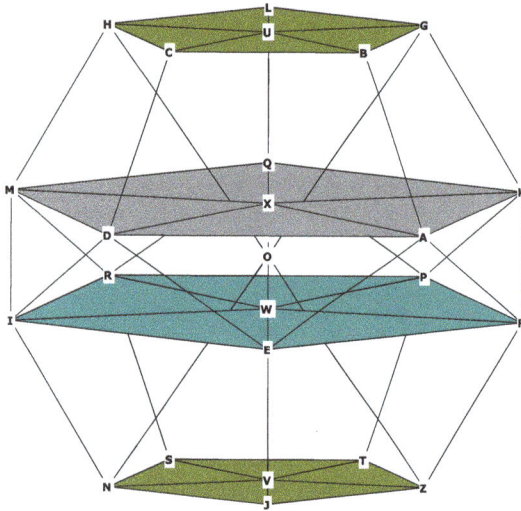

Figure 11.12: Figure 11.8, repeated

In Figure 11.12 we can see that $\overline{UH}$ on the top plane is the distance from the pentagon center to a vertex.

On plane ADMQK, $\overline{XM}$ is parallel to $\overline{UH}$ and is also the distance from that pentagon center to a vertex. $\overline{UH}$ is connected to $\overline{XM}$ by $\overline{HM}$, a side of the pentagon face CDIMH.

So we have a quadrilateral UHMX, with $\overline{UH}$ parallel to $\overline{XM}$. From here we can derive $\overline{UX}$, the distance between the two planes.

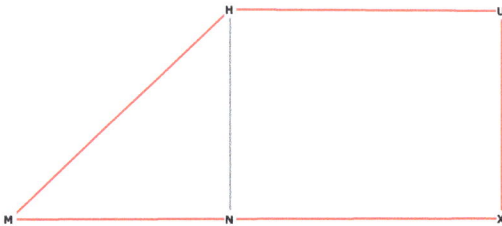

Figure 11.13: Dodecahedron planar distance $\overline{UX}$.

We know from Chapter 8 *Construction of the Pentagon Part II* that the distance from the center of pentagon to a vertex $= \frac{\Phi}{\sqrt{\Phi^2+1}}s$.

Therefore $\overline{UH} = \frac{\Phi}{\sqrt{\Phi^2+1}}s$.

The side of the large pentagon ADMQK in Figure 11.12 is, as we have seen, a diagonal of a dodecahedron face and so the side of the large pentagon is $\Phi \cdot s$. Therefore, $\overline{XM} = \frac{\Phi^2}{\sqrt{\Phi^2+1}}s$.

Therefore, $\overline{XM} = \Phi\left(\overline{UH}\right)$, and $\overline{XM}$ is divided in Mean and Extreme Ratio at N (See Figure 11.13).

$$\overline{MN} = \overline{XM} - \overline{XN} = \frac{\Phi^2}{\sqrt{\Phi^2+1}} - \frac{\Phi}{\sqrt{\Phi^2+1}} = \frac{1}{\sqrt{\Phi^2+1}}.$$

$\overline{HM} = s$. Triangle $\triangle MNH$ is right by construction.

So

$$\overline{HN}^2 = \overline{HM}^2 - \overline{MN}^2$$

$$= s^2 - \frac{1}{\Phi^2 + 1}\, s^2 = \frac{\Phi^2 + 1 - 1}{\Phi^2 + 1}\, s^2 = \frac{\Phi^2}{\Phi^2 + 1}\, s^2.$$

$$\overline{HN} = \overline{UX} = \frac{\Phi}{\sqrt{\Phi^2 + 1}}\, s. \tag{11.14}$$

Notice that $\overline{UH} = \overline{UX}$.

The dodecahedron is designed such that the distance to the two large pentagonal planes from the top or bottom faces is exactly equal to the distance between the center and a vertex of any of the faces of the dodecahedron.

This relationship is precisely what we saw in the icosahedron! That makes sense because the two are duals of each other.

The difference is that the dodecahedron is entirely pentagonal, both internally, and externally, on its faces.

11.8. Distance between the Two Large Pentagonal Planes

What is the distance $\overline{XW}$ between the two large pentagonal planes ADMQK and FEIRP? Figure 11.12 is misleading, it looks like the distance must be $\overline{MI}$ or $\overline{KF}$, the dodecahedron side, but it isn't.

We already have enough information to establish this distance.

$$\overline{UX} = \overline{VW}. \quad \overline{UV} = 2 \cdot \text{height of any pyramid} = \frac{2\Phi^3}{2\sqrt{\Phi^2 + 1}}\, s = \frac{\Phi^3}{\sqrt{\Phi^2 + 1}}\, s.$$

So

$$\overline{XW} = \overline{UV} - 2\left(\overline{UX}\right) = \frac{\Phi^3}{\sqrt{\Phi^2 + 1}} - \frac{2\Phi}{\sqrt{\Phi^2 + 1}} = \frac{\Phi^3 - 2\Phi}{\sqrt{\Phi^2 + 1}}\, s.$$

$$\overline{XW} = \frac{1}{\sqrt{\Phi^2 + 1}}\, s. \tag{11.15}$$

Notice that $\frac{\overline{UX}}{\overline{XW}} = \Phi$.

$\overline{UW}$ is divided in Mean and Extreme Ratio at X.

$\frac{\overline{XV}}{\overline{WV}} = \Phi$. $\overline{XV}$ is divided in Mean and Extreme Ratio at W.

A chart of these planar distances along the diameter of the dodecahedron, as we did with the icosahedron, can be seen in the Reference section at the end of this chapter on page 88 (refer to Figure 11.12).

11.9. Distance from Top Plane to Circumsphere

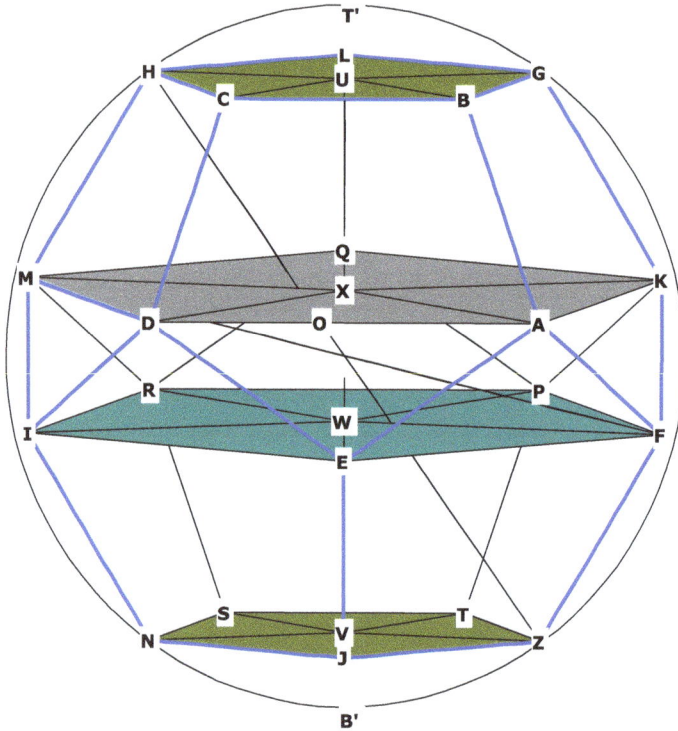

Figure 11.14: Dodecahedron in circumsphere

What is the distance from U to the top of the sphere, and from V to the bottom of the sphere?

The diameter of the enclosing sphere is $\overline{HZ}$, or $\sqrt{3}\Phi\, s$.

Let T' be the top of the sphere and B' be the bottom of the sphere. Refer to Figure 11.14.

If the radius is $\frac{\sqrt{3}\Phi}{2}$ and the distance $\overline{OU}$ is $\frac{\Phi^3}{2\sqrt{\Phi^2+1}}$, then

$$\overline{UT'} = \overline{VB'} = \frac{\sqrt{3}\Phi}{2} - \frac{\Phi^3}{2\sqrt{\Phi^2+1}} = 0.287742174\, s. \qquad (11.16)$$

Finally, let us demonstrate how the dodecahedron may be constructed from the interlocking vertices of 5 tetrahedra. We have already seen how the cube fits inside the dodecahedron, and how 2 interlocking tetrahedron may be formed from the diagonals of the cube. As Buckminster Fuller has pointed out, however, the cube and the dodecahedron are structurally unsound unless bolstered by the additional struts supplied by the tetrahedron. Fuller concludes logically that the tetrahedron is the basic building block of Universe; yet it is the dodecahedron that provides the blueprint and forms the structure for the interlocking tetrahedrons. The dodecahedron unites the geometry of crystals and lattices (root 2 and root 3) with the geometry of Phi (root 5), found in the biology of organic life.

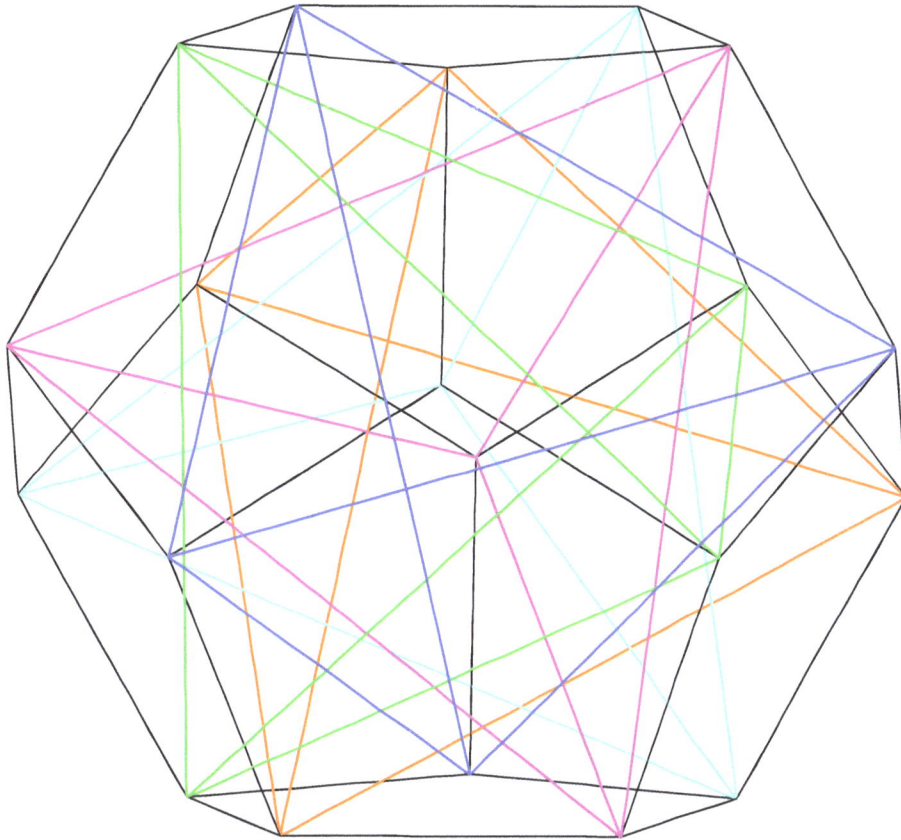

Figure 11.15: Dodecahedron from 5 tetrahedrons

11.10. Conclusion

The dodecahedron is entirely pentagonal, consisting of the $\sqrt{5}$ geometry of Phi. Yet it contains the $\sqrt{3}$ and $\sqrt{2}$ geometry of the cube, tetrahedron, and octahedron.

Remarkably, the sides of the cube are Φ times the side of the dodecahedron, because the cube side is the diagonal of a pentagonal face. Here is the key to the relationship of the first three Regular Solids and the much more complex icosahedron and dodecahedron.

Later on in this book we will discover a remarkable polyhedron that defines the relationship and provides the proper nesting for all 5 Platonic Solids directly on its vertices. If a polyhedron could be called exciting, this one is IT! If you can't wait, go to the last chapter of the book.

11.11. Dodecahedron Reference Tables

Table 11.1: Pentagonal planes of dodecahedron, relative to the side of the dodecahedron. graph reads vertically, by column

U		U	U
$\frac{\Phi}{\sqrt{\Phi^2+1}} = 0.85065$			
X	X		$\frac{\Phi^3}{2\sqrt{\Phi^2+1}} = 1.11352$
$\frac{1}{2\sqrt{\Phi^2+1}} = 0.2629$		$\frac{\Phi^2}{\sqrt{\Phi^2+1}} = 1.3764$	
O	$\frac{1}{\sqrt{\Phi^2+1}} = 0.5257$		O
$\frac{1}{2\sqrt{\Phi^2+1}} = 0.2629$			
W	W	W	$\frac{\Phi^3}{2\sqrt{\Phi^2+1}} = 1.11352$
$\frac{\Phi}{\sqrt{\Phi^2+1}} = 0.85065$			
V			V

Table 11.2: Here is a table of these relationships, letting $\overline{UX} = 1$: graph reads vertically, by column

U		U	U
1			
X	X		$\frac{\Phi^2}{2}$
$\frac{1}{2\Phi}$		Φ	
O	$\frac{1}{\Phi}$		O
$\frac{1}{2\Phi}$			
W	W	W	$\frac{\Phi^2}{2}$
1			
V			V

Table 11.3: Volume and Surface Area

Volume (edge)	Volume in Unit Sphere	Surface Area (edge)	Surface Area in Unit Sphere
7.663118963 s^3	2.785163863 r^3	20.64572881 s^2	10.51462224 r^2

Table 11.4: Angles

Central Angle:	Dihedral Angle:	Surface Angle:
41.81031488°	116.5650512°	108°

Table 11.5: Centroid Distances

Centroid To Vertex:	Centroid To Mid-edge:	Centroid To Mid-face:
1.0 r	0.934172359 r	0.794654473 r
1.401258539 s	1.309016995 s	1.113516365 s

Table 11.6: Side to Radius

Side / radius
0.713644179

Part IV

In Part IV we study the Cube Octahedron, the Star Tetrahedron, the Rhombic Dodecahedron, the Icosa-dodecahedron, and the amazing Rhombic Triacontahedron

Chapter 12

The Cube Octahedron

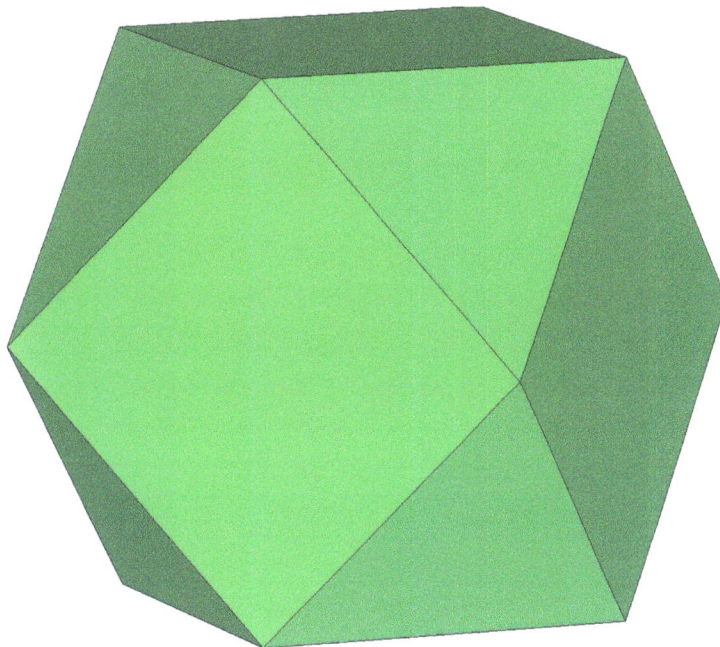

Figure 12.1: Cubeoctahedron

The cubeoctahedron is sometimes called the vector equilibrium. That is what Buckminster Fuller called it.

It has 12 vertices, 14 faces, and 24 edges. It is made by cutting off the corners of a cube.

Notice that in Figure 12.2 the triangular face of the cubeoctahedron IJK is formed by cutting off the corner of the cube G, and that the square face of the cube octahedron NPJI is formed when the four corners of the cube H,E,F, and G are cut off.

The cubeoctahedron has 8 triangular faces and 6 square faces. Figure 12.2 shows three of the square faces and four of the triangular faces.

There are 6 square faces on the cube octahedron, one for each face of the cube.

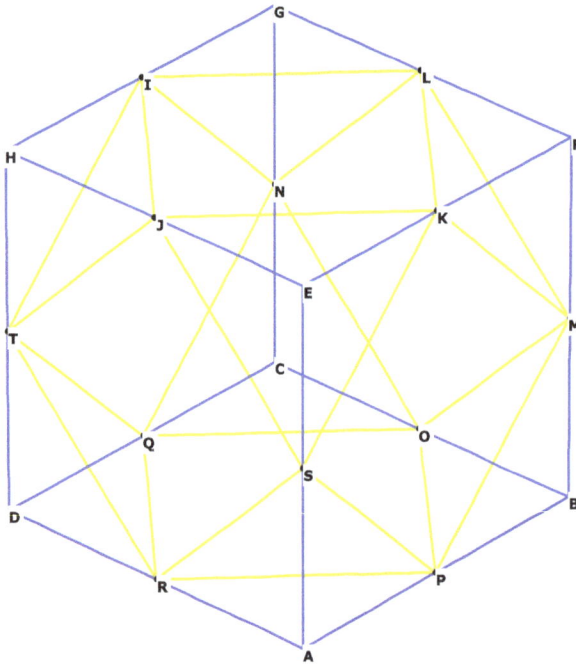

Figure 12.2: Showing the cube octahedron (yellow) inside the cube (blue).

There are 8 triangular faces on the cube octahedron, one for each vertex of the cube.

This polyhedron has the fascinating property that the radius of the enclosing sphere, which touches all 12 vertices, is exactly equal to the length of all of the sides of the cube octahedron. Unfortunately, our 2 dimensional perspective cannot accurately capture the cube octahedron in true perspective, but if you build a model of one you'll see its true.

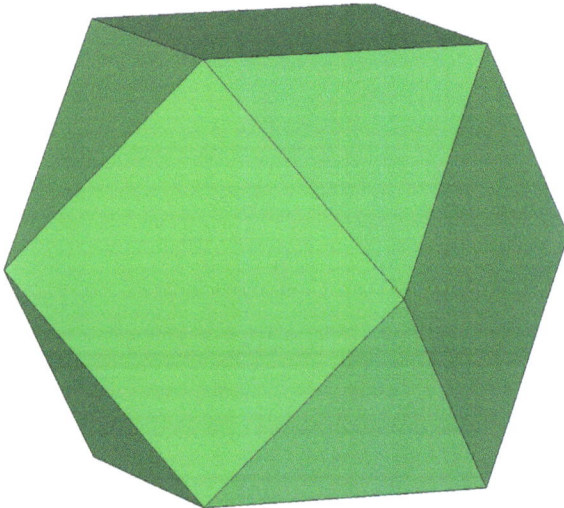

Figure 12.3: Cubeoctahedron showing triangular and square faces

In Figure 12.3 you can see some of the rays coming out from O, for example $\overline{OB}$ and $\overline{OA}$. All of these rays are equal in length to the sides, for example, $\overline{GL}$ and $\overline{LC}$. This means that all of the rays branching out from the centroid, O, do so at 60 degree angles, forming four "great circle" hexagonal planes on the outside of the figure.

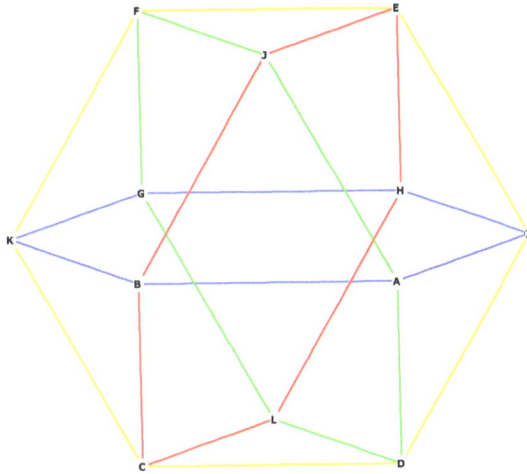

Figure 12.4: Showing the 4 "great circle" hexagons of the cubeoctahedron

The cube octahedron, if you observe Figure 12.3 closely, can be seen to be composed of 8 tetrahedrons and 6 half-octahedrons. OGLH and OJAB, for example, are tetrahedrons. The half-octahedrons are formed from the square planes, for example OEIAJ and OCKGL.

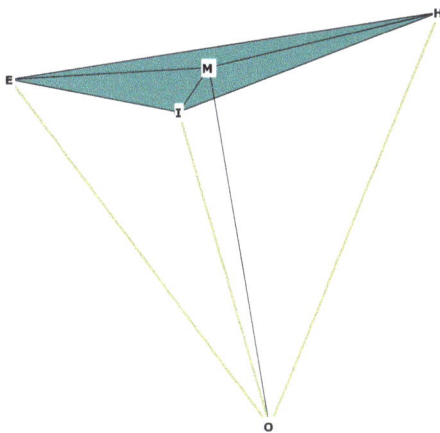

Figure 12.5: Tetrahedron within the cubeoctahedron

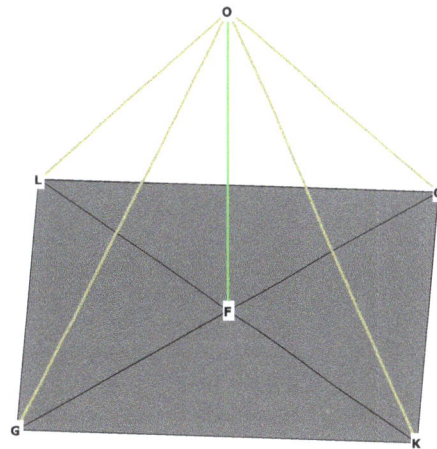

Figure 12.6: Half-octahedron within the cubeoctahedron

The 12 vertices of the cube octahedron can also be considered to be composed of 3 orthogonal squares centered around O, the sides of which cross through the square faces of the cube octahedron as the diagonals of the squares:

Note how the sides of the squares in Figure 12.7 are the diagonals of the square faces of the cube octahedron. Notice that in Figure 12.8 below, the intersection points of the squares form the 6 vertices of an octahedron (in purple).

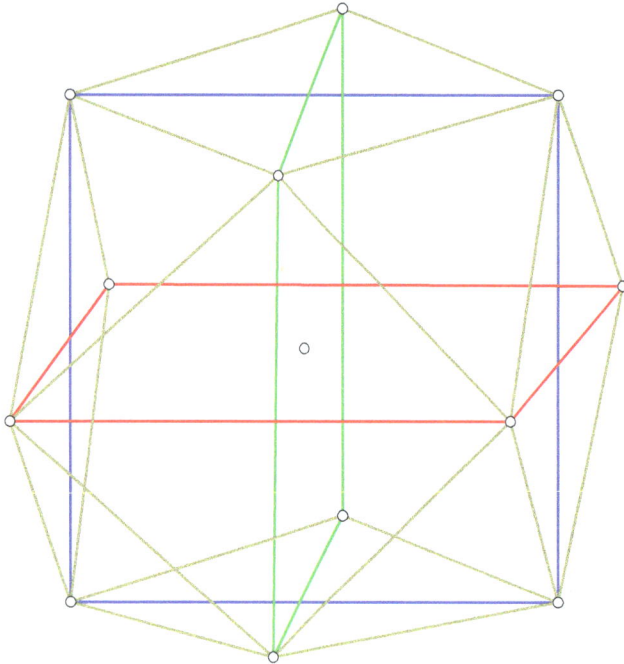

Figure 12.7: Three interlocking squares of the cubeoctahedron

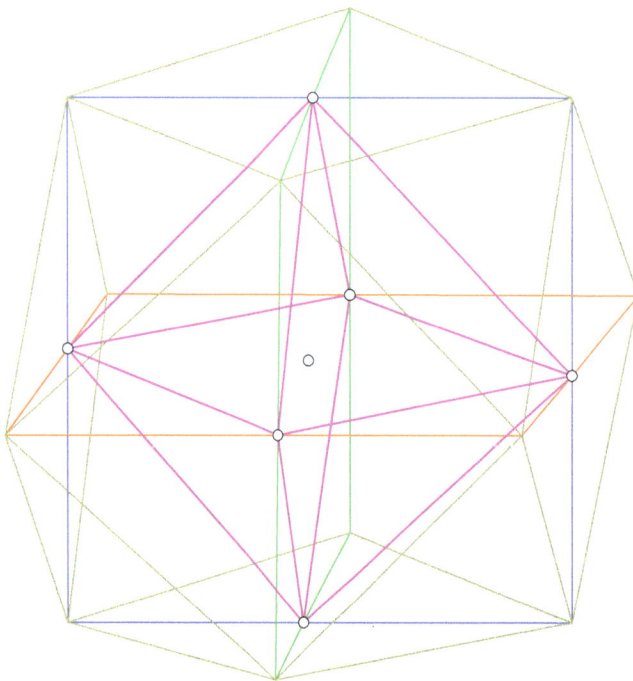

Figure 12.8: Octahedron inside cubeoctahedron

12.1. Radius to Side in Cubeoctahedron

What is the ratio of the radius of the enclosing sphere around the cube octahedron, and the side of the cube octahedron?

$$r = s. \tag{12.1}$$

All line segments from the centroid to any vertex are identical in length to the edge length's of the cube octahedron.

12.2. Volume of the Cubeoctahedron

The volume of the cube octahedron is hereinafter referred to as c.o.

We use the pyramid method as usual.

$V = \frac{1}{3}$ (area of base) (pyramid height).

We have 8 triangular faces and 6 square faces.

From *The Equilateral Triangle* we know that the area of the triangular face is $\frac{\sqrt{3}}{4} s^2$.

The area of the square face is just $s \cdot s = s^2$.

We need to find the height of the square pyramid and the height of the triangular pyramid.

In Figure 12.5, the triangular pyramid, or tetrahedron, is OEHI, with height $\overline{OM}$. Triangle OMH is right, the line $\overline{OM}$ being perpendicular to the plane EHI at M. The height of the tetrahedral pyramid is known, from *Tetrahedron*, as $\frac{\sqrt{2}}{\sqrt{3}} s$.

In Figure 12.6, F is the center of the square face LCKG, and $\overline{OF}$ is the pyramid height. The triangle OFL is right, the line $\overline{OF}$ being perpendicular to the plane LCKG at F. $\overline{OL}$ is just the radius of the enclosing sphere, which, in the c.o., is the same as the c.o. side, s.

$\overline{LF}$ is one half the diagonal of the square LCKG, or $\frac{1}{\sqrt{2}} s$, because the diagonal of a square is always $\sqrt{2}$ times the side of the square.

Therefore we can write for the pyramid height $\overline{OF}$, $\overline{OF}^2 = \overline{OL}^2 - \overline{LF}^2 = 1s^2 - \frac{1}{2} s^2 = \frac{1}{2} s^2$.

$\overline{OF} = \frac{1}{\sqrt{2}} s$. Note that this distance is identical to $\overline{LF}$.

$$Volume_{\text{1 triangular pyramid}} = \frac{1}{3} \left(\frac{\sqrt{3}}{4} s^2 \right) \left(\frac{\sqrt{2}}{\sqrt{3}} s \right) = \frac{\sqrt{2}}{12} s^3.$$

$$Volume_{\text{all 8 triangular pyramids}} = 8 \left(\frac{\sqrt{2}}{12} s^3 \right) = \frac{2\sqrt{2}}{3} s^3.$$

$$Volume_{\text{1 square pyramid}} = \frac{1}{3} s^2 \left(\frac{1}{\sqrt{2}} s \right) = \frac{1}{3\sqrt{2}} s^3.$$

$$Volume_{\text{all 6 square pyramids}} = 6 \left(\frac{1}{3\sqrt{2}} s^3 \right) = \frac{2}{\sqrt{2}} s^3.$$

Therefore,

$$Total\ Volume_{\text{cube octahedron}} = \frac{2\sqrt{2}}{3} s^3 + \frac{2}{\sqrt{2}} s^3 = \frac{5\sqrt{2}}{3} s^3$$

$$= 2.357022604\, s^3 = 2.357022604\, r^3. \tag{12.2}$$

This figure is larger than for the cube, but why? The volume of the cube in the unit sphere is only $1.539600718\, r^3$! The reason lies in the fact that the cubeoctahedron fits more snugly within the unit sphere than does the cube.

The model on my desk shows the cube octahedron inside the cube, as in Figure 12.2. Here, the side of the cube octahedron is $\frac{1}{\sqrt{2}}$ (side of cube.) This can be determined by an inspection of triangle $\triangle GIJ$ in Figure 12.2.

$\overline{GI} = \overline{GJ} =$ one-half the side of the cube. Angle $\angle IGJ$ is right. Therefore, $\overline{GJ}$, the side of the cube octahedron, $= \sqrt{\frac{1}{4} + \frac{1}{4}} = \frac{1}{\sqrt{2}} \cdot$ side of cube.

Calculating the volume of the cubeoctahedron in terms of the side of the cube as in Figure 12.2, we write

$$Volume_{\text{cubeoctahedron}} = \frac{5\sqrt{2}}{3} \left(\frac{1}{\sqrt{2}} \right)^3 = \frac{5\sqrt{2}}{3} \left(\frac{1}{2\sqrt{2}} \right) = \frac{5}{6} soc^3, \tag{12.3}$$

where soc is the side of the cube.

Therefore the volume of the cubeoctahedron is 5/6ths the volume of the cube when the cubeoctahedron is sitting inside the cube.

There are 8 small tetrahedrons which represent the volume of the cube octahedron that have been "cut out" of the cube (See Figure 12.2 and the tetrahedron GKIJ, for example). The volume of each of these tetrahedrons must then be one–eighth of 1/6, the difference between the volume of the cube and the volume of the cubeoctahedron. Therefore the volume of each small tetrahedron $= \frac{1}{8} \cdot \frac{1}{6} = \frac{1}{48} soc^3$, where soc is the side of the cube..

12.3. Surface Area of Cubeoctahedron

The surface area is 8 (area of triangular face) + 6 (area of square face) =

$$8 \left(\frac{\sqrt{3}}{4} s^2 \right) + 6 s^2 = (6 + 2\sqrt{3}) s^2 = 9.464101615 s^2. \tag{12.4}$$

12.4. Central angle of the Cubeoctahedron

Each of the internal angles of the c.o. arc formed from any of the 4 hexagonal planes which surround the centroid (see Figure 12.4, p. 95).

$$\text{Therefore the central angle of the cube octahedron} = 60°. \qquad (12.5)$$

12.5. Surface Angles of the Cubeoctahedron

There are two surface angles of the cubeoctahedron, one being the 60° angle of the triangular faces, the other being the 90°angle of the square faces.

12.6. Dihedral Angle of the Cubeoctahedron

What is the dihedral angle of the cube octahedron? This angle is the intersection between a square face and a triangular face. If you sit the c.o. on one of its square faces you can see the view in Figure 12.9.

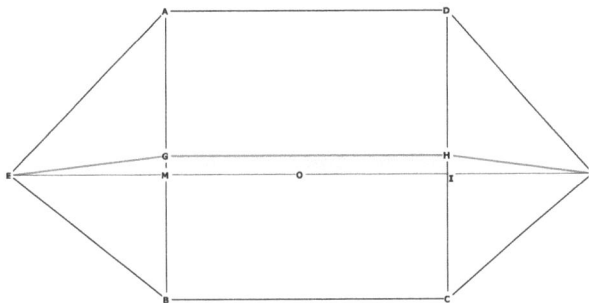

Figure 12.9: Illustrating the dihedral angle of the cube octahedron – top view

I have drawn the line EGHF to illustrate the dihedral angle between the two triangular faces and the square face ABCD. The point G is directly above the point M. O is the centroid of the c.o. M bisects $\overline{EO}$.

The triangle $\triangle GME$ is right by construction. The angle $\angle MGH$ is right by construction. The angle $\angle EGH$ is the dihedral angle. If we can find the angle $\angle EGM$, all we have to do then is add 90° to it ($\angle MGH$), and we have the dihedral angle.

$\overline{OE} = \overline{OF} =$ side of c.o., or s. $\overline{EM} = \frac{1}{2}\left(\overline{OE}\right)$.

$\overline{EG}$ is the height of a triangular face, which we know from *The Equilateral Triangle* $= \frac{\sqrt{3}}{2}s$.

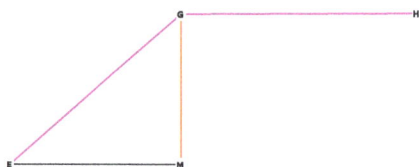

Figure 12.10: Showing that GME and MGH are right angles

So we can write:

$$\sin(\angle EGM) = \frac{\overline{EM}}{\overline{EG}} = \frac{\frac{1}{2}}{\frac{\sqrt{3}}{2}} = \frac{1}{\sqrt{3}}.$$

$$\arcsin\left(\frac{1}{\sqrt{3}}\right) = 35.26438968°.$$

$$\angle EGH = 90° + 35.26438968° = 125.26438968°.$$

$$\text{Dihedral angle} = 125.26438968°. \tag{12.6}$$

Of course, the dihedral angle between the square faces is $90°$.

12.7. Centroid Distances

The distance from the centroid to each vertex is $s = r$.

The distance from the centroid to any mid-edge is just the height of an equilateral triangle (remember that the centroid is surrounded by 4 hexagons (see Figure 12.3). This distance is $\frac{\sqrt{3}}{2}\,s$.

The distance from the centroid to a triangular mid-face is just the height of the tetrahedron (see Figure 12.4), which we know from above, as $\frac{\sqrt{2}}{\sqrt{3}}\,s$.

The distance from the centroid to a square mid-face is just the height of the half-octahedron $\overline{OF}$ (see Figure 12.5), which we know from above, is $\frac{1}{\sqrt{2}}\,s$.

The relationship between these distances is 1, 0.866025404, 0.816496581, 0.707106781

12.8. Cubeoctahedron Reference Tables

Table 12.1: Volume and Surface Area

Volume (edge)	Volume in Unit Sphere	Surface Area (edge)	Surface Area in Unit Sphere
$2.357022604\ s^3$	$2.357022604\ r^3$	$9.464101615\ s^2$	$9.464101615\ r^2$

Table 12.2: Angles

| Central Angle | Dihedral Angles | | Surface Angles | |
	between square & triangular faces	between square faces	triangular faces	square faces
$60°$	$125.26438968°$	$90°$	$60°$	$90°$

Table 12.3: Centroid Distances

Centroid To Vertex	Centroid To Mid-edge	Centroid To Mid-triangle Face	Centroid To Mid-square Face
$1.0\ r$	$0.866025404\ r$	$0.816496581\ r$	$0.707106781\ r$
$1.0\ s$	$0.866025404\ s$	$0.816496581\ s$	$0.707106781\ s$

Table 12.4: Side to Radius

Side / radius
1.0

Chapter 13

The Star Tetrahedron

13.1. Overview

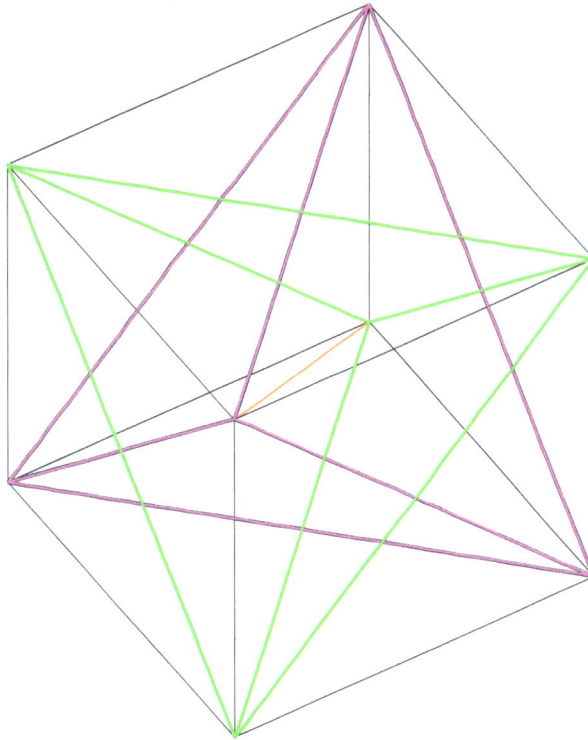

Figure 13.1: The star tetrahedron in the cube

Figure 13.1 shows the cube in thin black, and two interlocking tetrahedrons, one in purple, the other in green.

To visualize this, look at the orange line and imagine that the purple tetrahedron comes out of the page and the green tetrahedron goes back into the page.

The two interlocking tetrahedrons each intersect the other by bisecting each others' sides, and form the vertices of an octahedron. This becomes much clearer when you build a 3D model.

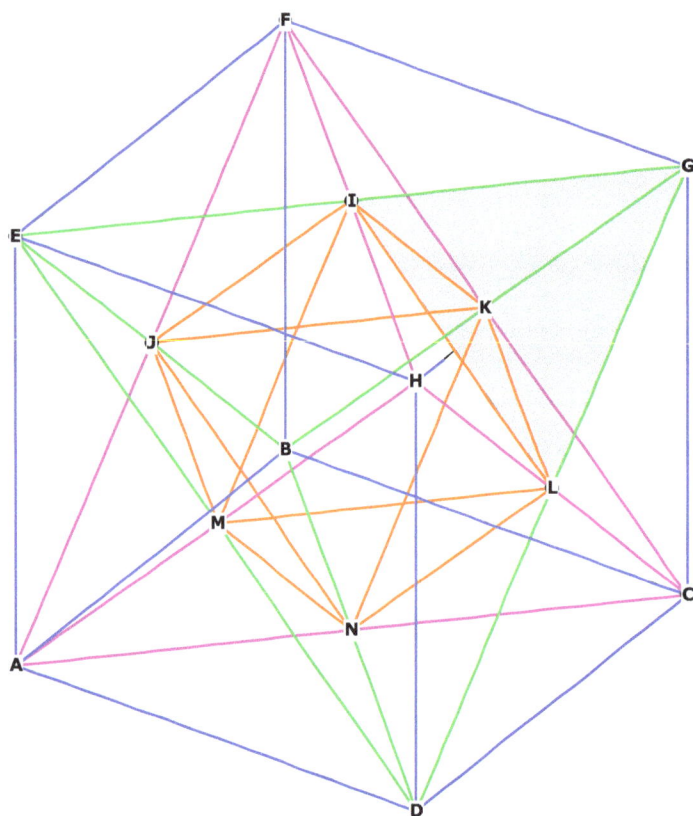

Figure 13.2: Showing one of the eight smaller tetrahedra that stick out from the star tetrahedron in the cube. the bottom plane of each of these tetrahedra form one of the faces of an octahedron inside the star tetrahedron

Figure 13.2 shows that the interlocking tetrahedra bisect the sides of each other. $\overline{BG}$ bisects $\overline{CA}$ at H, $\overline{FG}$ bisects $\overline{AE}$ at J, $\overline{DG}$ bisects $\overline{CE}$ at I. There are 12 interlocking edges, 6 for each tetrahedron.

The intersection of the two interlocking tetrahedrons forms an octahedron plus 8 smaller tetrahedron's that stick out from the octahedron, as shown in Figure 13.2.

13.2. Volume of Star Tetrahedron

The volume of the star tetrahedron is the volume of the octahedron + volume of each small tetrahedron, so we write:

$$Volume_{\text{star tetrahedron}} = Volume_{\text{octahedron}} + Volume_{\text{8 small tetrahedrons}}$$

Let cs = side of the cube = 1. Then $Volume_{\text{cube}} = 1^3 = 1$.

The length of the edges of the interlocking tetrahedrons are each $\sqrt{2}\cdot$ side of the cube, because each of them is a diagonal of a square face of the cube.

The edges of each of the small tetrahedrons are exactly one-half the side of the large tetrahedron, because each of the large tetrahedrons bisects the other in order to form the interior octahedron. That is the magic of the Platonic Solids!

So the sides of the small tetrahedron $= \frac{\sqrt{2}}{2}cs$.

The sides of the interior octahedron are also $\frac{\sqrt{2}}{2}cs$, because the vertices of the octahedron bisect the edges of the interlocking tetrahedrons. This becomes clear when you build a three dimensional model.

We know from Chapter 2 *Octahedron* that the volume of an Octahedron is $\frac{\sqrt{2}}{3}\cdot$side of octahedron, or os. We need to convert this to the side of the cube:

$$Volume_{\text{octahedron}} = \frac{\sqrt{2}}{3}\left(\frac{\sqrt{2}}{2}cs\right)^3 = \frac{\sqrt{2}}{3}\left(\frac{2\sqrt{2}}{8}cs^3\right) = \frac{1}{6}cs^3.$$

We know from Chapter 1 *Tetrahedron* that the volume of a Tetrahedron is $\frac{1}{6\sqrt{2}}\left(ts^3\right)$, where $ts =$ side of tetrahedron. It turns out that the edge lengths of each of our 8 tetrahedrons is equal to the lengths of the sides of the octahedron. So we may write

$$Volume_{\text{small tetrahedron}} = \frac{1}{6\sqrt{2}}\left(\frac{\sqrt{2}}{2}cs\right)^3 = \frac{1}{6\sqrt{2}}\left(\frac{\sqrt{2}}{4}cs^3\right) = \frac{1}{24}cs^3.$$

$$Volume_{\text{total, 8 small tetrahedrons}} = 8\left(\frac{1}{24}cs^3\right) = \frac{1}{3}cs^3.$$

$$Volume_{\text{total star tetrahedron}} = \frac{1}{6}cs^3 + \frac{1}{3}cs^3 = \frac{1}{2}cs^3. \tag{13.1}$$

Even though the volume of each large tetrahedron in the star tetrahedron is exactly one-third of the cube, the star tetrahedron only occupies one half the volume of the cube.

13.3. Volume of a Left-over Solid in the Cube

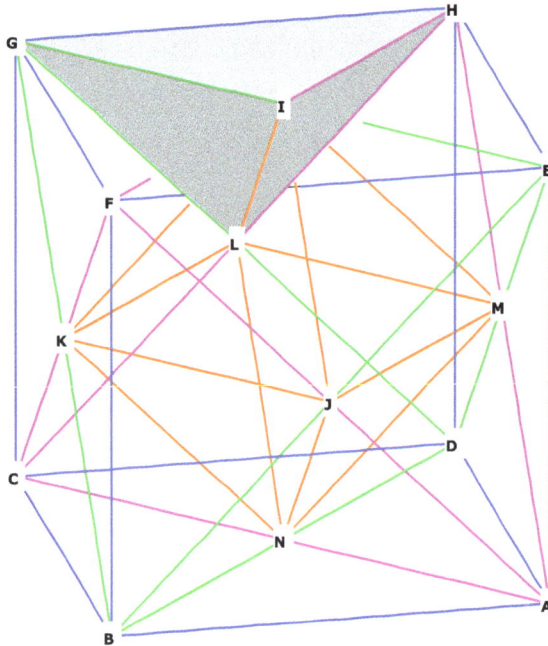

Figure 13.3: Showing the 8 congruent tetrahedra (for example, GLHI) representing the left over space in the cube not included in the star tetrahedron volume.

Figure 13.3 shows the 8 congruent tetrahedral solids representing the left over space in the cube not included in the star tetrahedron volume. These are different from the tetrahedra shown in Figure 13.2, which are actually part of the star tetrahedron itself. If you build a 3-D model you will see the distinction immediately.

Since the left over space in the cube $= \frac{1}{2} cs^3$,

$$Volume_{\text{one left over solid}} = \frac{1}{8}\left(\frac{1}{2} cs^3\right) = \frac{1}{16} cs^3. \tag{13.2}$$

These solids have 4 vertices, 4 faces, and 6 edges, thus fulfilling the Euler requirement that Faces + Vertices = Edges + 2.

13.4. The Star Tetrahedron in the Sphere

The sphere enclosing the star tetrahedron is the same sphere that encloses the cube.

The diameter of the enclosing sphere is the distance $\overline{ED}$ marked in orange in Figure 13.4.

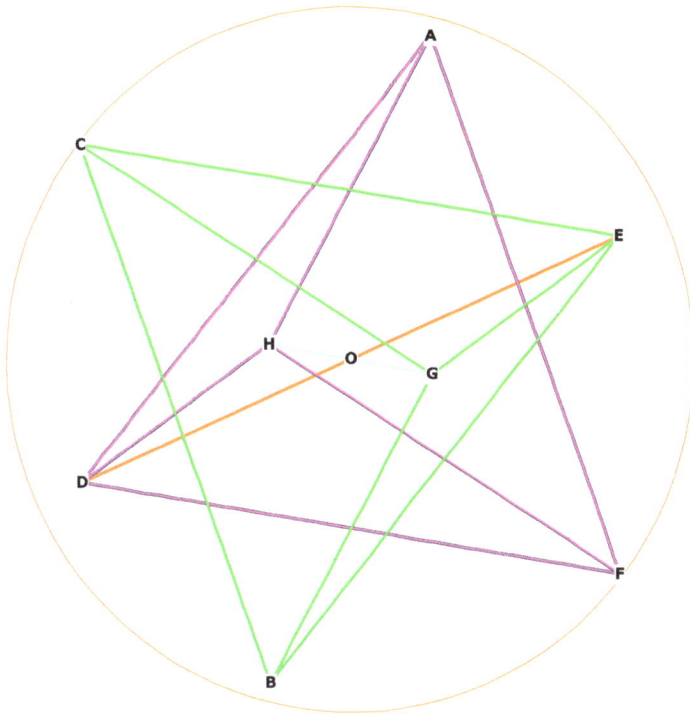

Figure 13.4: Star Tetrahedron in the sphere

All 8 points of the star tetrahedron (A, B, C, D, E, F, G, H) touch the surface of the enclosing sphere. Note that the diameter of the enclosing sphere is the same diagonal that goes through the center of the cube. In Figure 13.4 this is marked as $\overline{ED}$, with the centroid at O. From *Cube* we know that the length of this diagonal, and the diameter of the enclosing sphere, is $\sqrt{3}$. (This calculation is a simple one, $\overline{ED}$ being the hypotenuse of the right triangle $\triangle EDF$, which is apparent in Figure 13.5. Since $\overline{DF}$ is the diagonal of the cube with sides $= 1$, $\overline{DF} = \sqrt{2}$. $\overline{FE} = 1$, because it's the side of the cube, so by the Pythagorean Theorem, $\overline{ED} = \sqrt{3}$).

There are 4 axes of rotation for the star tetrahedron; $\overline{AB}, \overline{CF}, \overline{DE}$, and $\overline{GH}$.

Notice the planes CGE in green and HDF in purple; both planes go into and out of the screen or page.

13.5. Distance between Planes CGE and HDF

What is the distance between these planes? This distance will be the line $\overline{JOL}$ in Figure 13.5. J lies at the center of CGE, and L lies at the center of HDF. O is the centroid. By construction, both planes are parallel to each other.

It turns out that if you have a model of the star tetrahedron you can easily see that the distance between the two planes, $\overline{JOL}$, is the height of one of the smaller tetrahedrons from the previous section.

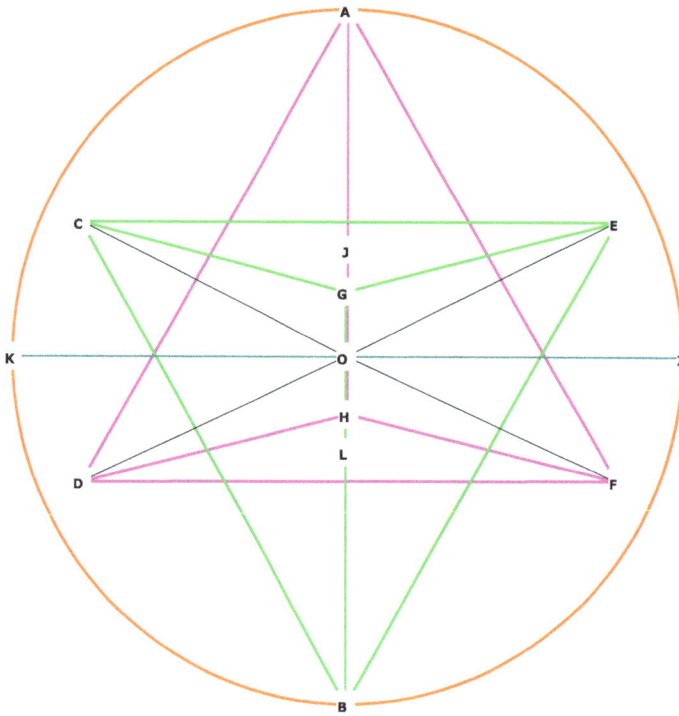

Figure 13.5: Another view of star tetrahedron in the sphere

We know that the height of a tetrahedron is

$$\frac{\sqrt{2}}{\sqrt{3}} \text{ (side of tetrahedron)} = \frac{\sqrt{2}}{\sqrt{3}} \left(\frac{\sqrt{2}}{2}\right) \text{ (side of cube)} = \frac{1}{\sqrt{3}} \text{ (side of cube)}.$$

$$\overline{JOL} = \frac{1}{\sqrt{3}} cs. \tag{13.3}$$

This distance is exactly one-third of the diameter of the enclosing sphere.

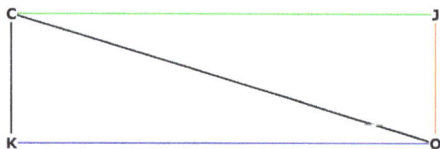

Figure 13.6: Showing $\overline{JO}$, the distance between the plane CGE and the centroid O

13.6. Angle between the xy plane at the Origin and the Planes DGE and HDF

This will be $\angle IOE$, for example, in Figure 13.5.

$\overline{JO} = $ one-half of $\overline{JOL} = \frac{1}{2\sqrt{3}} cs.$

$\overline{CJ}$ represents the plane CGE and $\overline{OK}$ represents the plane DFH. J is the center of the plane CGE, and L is the center of the plane DFH.

$\overline{CO}$ is the radius of the enclosing sphere, and is equal to $\frac{\sqrt{3}}{2}cs$.

Now we have $\overline{OJ}$ and $\overline{CO}$ and can get the angle $\angle COJ$:

$$\cos(\angle COJ) = \frac{\overline{OJ}}{\overline{OC}} = \frac{\frac{1}{2\sqrt{3}}}{\frac{\sqrt{3}}{2}} = \frac{1}{3}, \text{so}$$

$$\angle COJ = 70.52877937°, \text{and}$$

$$\angle COK = 19.47122063°. \tag{13.4}$$

Chapter 14

The Rhombic Dodecahedron

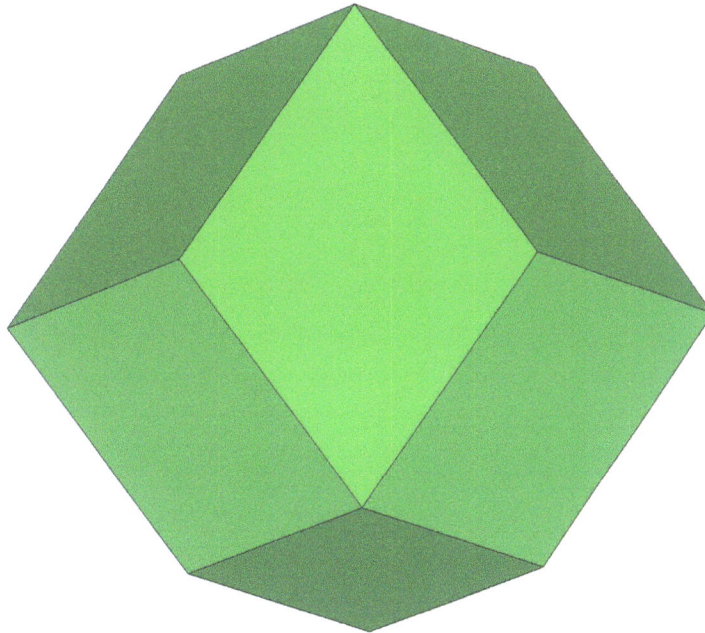

Figure 14.1: The Rhombic Dodecahedron

The rhombic dodecahedron is a very interesting polyhedron. It figures prominently in Buckminster Fuller's *Synergetics*.

It has 12 faces, 14 vertices, and 24 sides or edges.

14.1. Introduction

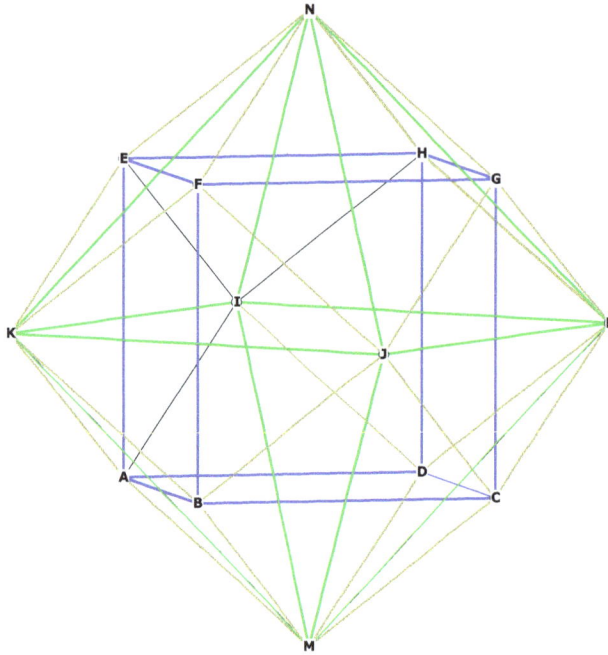

Figure 14.2: Rhombic dodecahedron showing inernal cube and octahedron

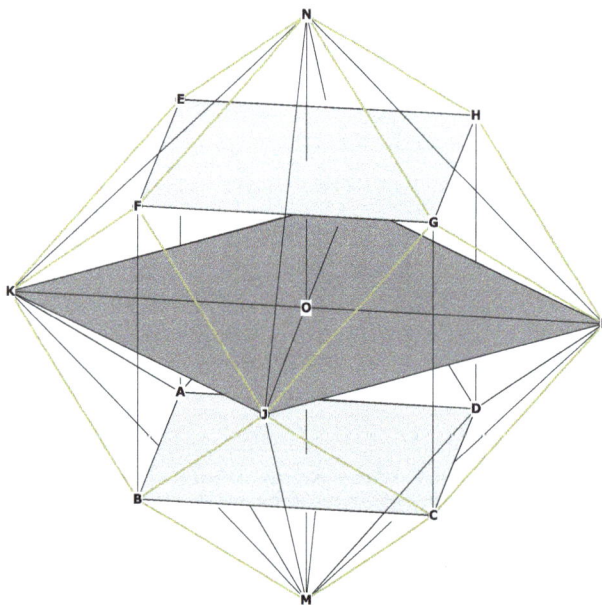

Figure 14.3: Internal planes of the rhombic dodecahedron

Figure 14.2 shows the edges of the rhombic dodecahedron (yellow), the internal octahedron (green) and cube (blue).

Notice in Figures 14.2 and 14.3 that the rhombic dodecahedron is composed of diamond faces (for example, NFKE).

The faces are called rhombuses, because they are equilateral parallelograms. In other words, they are square-sided figures with opposite edges parallel to one another.

The rhombic dodecahedron has 8 vertices in the middle that form a cube, (ABCD-EFGH), and the other 6 on the outside which form an octahedron (N-ILJK-M). It is possible to draw a sphere around the cube, and another, larger sphere, around the octahedron.

Unlike the five regular solids, therefore, not all of the vertices of the rhombic dodecahedron will touch one sphere. We will, in the course of this analysis, find the diameter and radius of each of these spheres.

In Figure 14.3, I have marked the top and bottom planes of the cube in light gray, and the square plane which serves as the base for the 2 face-bonded pyramids of the octahedron, in dark gray.

You may perceive three edges of the cube in this drawing, which connect the top and bottom planes; that is, $\overline{FB}, \overline{GC}, \overline{HD}$; and also some of the edges of the octahedron, the most visible of which are $\overline{NJ}, \overline{NK}, \overline{NL}, \overline{MK}, \overline{MJ}, \overline{ML}$. Notice that the edges of the octahedron bisect the diamond faces of the rhombic dodecahedron upon their long axis (for example, $\overline{NK}$ bisects the long axis of the face NEKF at the upper left). Note also that the short axis segments (that is, $\overline{EF}$) are the edges of a cube.

It is important to understand that the outer vertices of the rhombic dodecahedron form an octahedron, as this will be an important part of the analysis.

The rhombic dodecahedron (hereinafter, referred to as r.d.) is a semi-regular polyhedron, in that all of its edges are the same length, yet the angles of its faces differ. Because each face is a parallelogram, there are two distinct angles for each face, one which is bisected by the long axis, with an angle less than 90 degrees, the other bisected by the short axis, with an angle greater than 90 degrees. Of course these axes do not actually appear on the face of the polyhedron, I use them here for illustration.

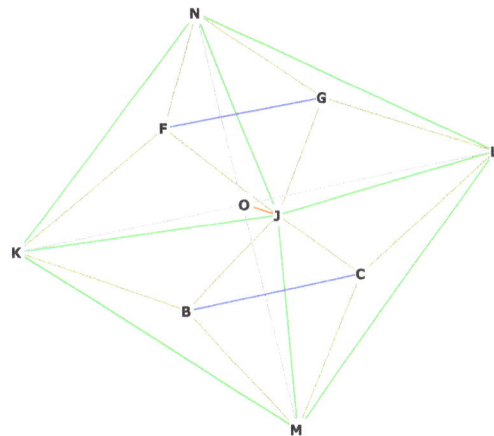

Figure 14.4: Rhombic dodecahedron internals

In Figure 14.4, the vertices of the octahedron are the long axis vertices of each face, in this case N, J, and M. The cube vertices are the short axis points, in this case, F, G, B, and C. This can also be seen by referring to Figure 14.3.

Figure 14.4 shows the general appearance of the r.d. The front four faces of the octahedron can be seen clearly here in green (NKJ, NJL, KJM and JLM). Figure 14.4 also shows how the short axis vertices of the r.d. (as F, G, B, C) come off the face of the octahedron. The long axis vertices (as in N, J) are vertices of the octahedron. O is the centroid and is beneath J.

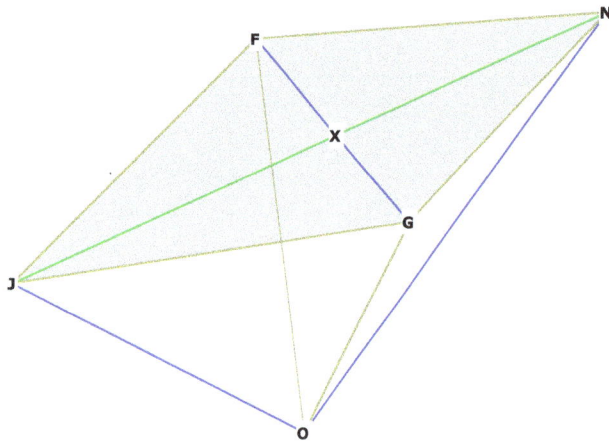

Figure 14.5: Long axis ($\overline{NJ}$) and short axis ($\overline{FG}$) of rhombic dodecahedron face

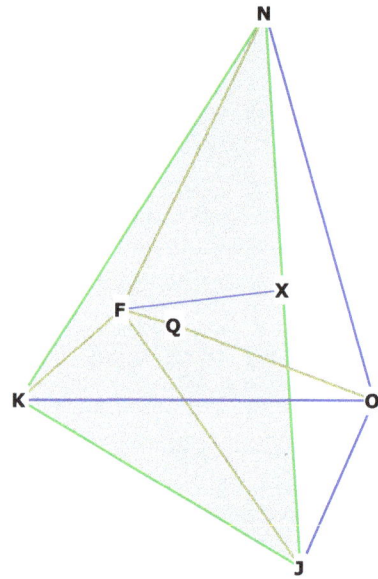

Figure 14.6: Point F, a vertex of the rhombic dodec-ahedron, off one of the octahedron faces NJK

The sides of the rhombic dodecahedron are JFNG, one of the sides of the octahedron is $\overline{JN}$. JON is the central angle of the long axis of the r.d. face, and FOG is the central angle of the short axis. O is the centroid of the r.d. and of the octahedron and cube within the r.d. When you build a 3D model of the r.d., it appears that $\overline{OF} = \overline{OG} = \overline{NF} = \overline{FJ} = \overline{JG} = \overline{NG}$ by construction; that is, the distance from the centroid O to any of the 6 short axis vertices of the r.d. faces (the vertices that make the cube, see Figure 14.3) are equal in length to the edges of the rhombic dodecahedron. We will prove this later on.

Figure 14.6 shows the point F, one of the vertices of the rhombic dodecahedron, off one of the faces of the octahedron, NJK. The centroid of the octahedron/r.d. is at O. Q is the center of the octahedral face NJK.

If you build a model of the rhombic dodecahedron with the Zometool, you will see at once that the sides of the r.d. come off any of the faces of the octahedron and meet above the center of the octahedron face (for example, at F). F is also the centroid of a tetrahedron with side length = side of the octahedron that can be formed from the face NJK of the octahedron (see Figure 14.7 below, the tetrahedron NJKZ). Tetrahedrons may be formed from any of the faces of the octahedron, and connected in the fashion of Fuller's Isotopic Vector Matrix.

$\overline{QF}$, in Figure 14.7, is the distance from the plane of the octahedron/tetrahedron face, to the centroid of any such tetrahedron. $\overline{XF}$ is the distance from the center of any of the diamond faces of the r.d. to a short axis vertex, in this case, F.

$\overline{OQ}$ is the distance from the centroid to the middle of the face of the octahedron. The centroid of the octahedron is also the centroid of the r.d. (O), which is built around the faces of the octahedron, as shown in Figures 14.2 and 14.3.

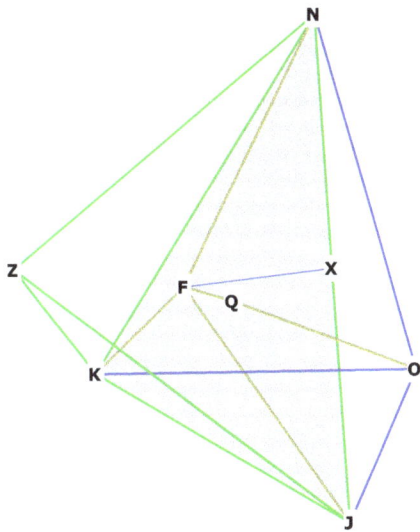

Figure 14.7: Building a tetrahedron off the octahedral face

14.2. Volume of the Rhombic Dodecahedron

Let's find the volume of the rhombic dodecahedron. As usual, we will use the pyramid method.

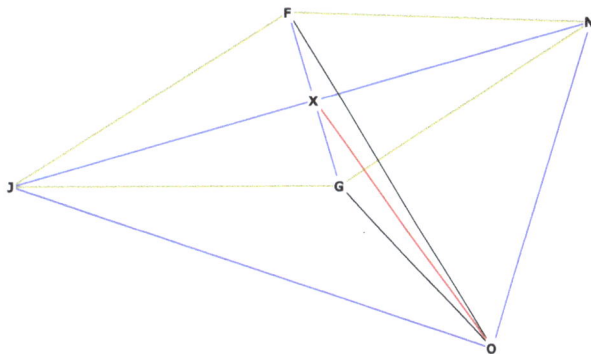

Figure 14.8: Rhombic dodecahedron pyramid

In Figure 14.8, $\angle JON$ is $90°$.

The height of any pyramid is $\frac{1}{3}$ (area of base) (height of pyramid).

We need to get the area of each diamond face.

In order to do this we need to take a rather extensive detour in which we will derive lots of interesting information about the rhombic dodecahedron.

14.2.1. RHOMBIC DODECAHEDRON INTERNALS

First off, we need to know the length of the side of the rhombic dodecahedron. In reference to what? We have an octahedron and a cube inside the r.d. We already have the length's of each of these sides, in relation to a sphere that encloses the octahedron and cube. The outer sphere of the r.d. is precisely that sphere that touches all 6 vertices of the octahedron, so we choose the edge of the octahedron as our reference point.

Recall from *Octahedron* that this relationship is:

$$r = \frac{1}{\sqrt{2}} \text{ (side of octahedron)}.$$

Observe that the long axis $\overline{NJ}$ in Figures 14.2 and 14.3 is the side of the octahedron. The distance $\overline{ON} = \overline{OJ}$ is the radius of the unit sphere that encloses the octahedron.

We have remarked previously that it is possible to build a tetrahedron off of any of the faces of the octahedron that lies within the r.d. as in Figure 14.9

The distance $\overline{FX}$ in Figure 14.9 is the distance from the centroid of the tetrahedron to the mid-edge of the tetrahedron. We know from Chapter 1 *Tetrahedron* that this distance is $\frac{1}{2\sqrt{2}}$ (side of the tetrahedron). Therefore, $\overline{FX} = \frac{1}{2\sqrt{2}}$ (side of the octahedron).

Now we can find the side of the r.d. in terms of the octahedron side.

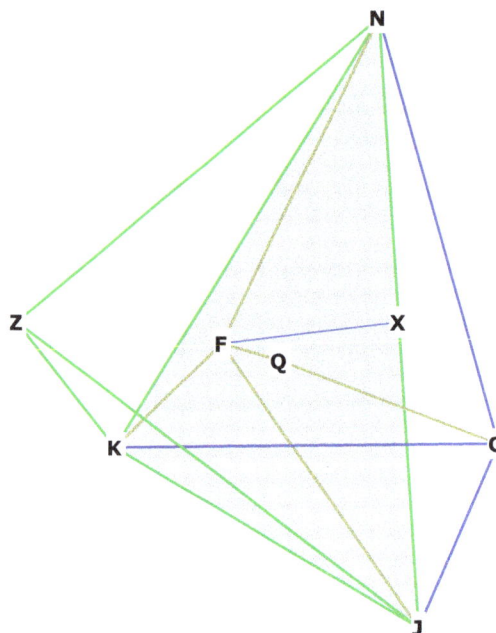

Figure 14.9: Building a tetrahedron off the internal octahedral face of the r.d.

14.2.2. SIDE OF RHOMBIC DODECAHEDRON IN TERMS OF THE OCTAHEDRON SIDE

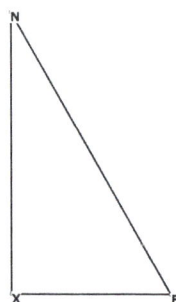

Figure 14.10

$\overline{NX}$ = one-half the side of the octahedron, or long-axis of the r.d. face (see Figure 14.10).

$\overline{NF}$ is the r.d. side.

We will refer to *os* as the side of the octahedron. *os* is equal in length to the side of a tetrahedron whose face is congruent to the octahedron face (See Figure 14.9).

rds will be the r.d. side.

$$\overline{NF}^2 = \overline{NX}^2 + \overline{FX}^2 = \frac{1}{4}os^2 + \frac{1}{8}os^2 = \frac{3}{8}\,os^2.$$

$$\overline{NF} = rds = \frac{\sqrt{3}}{2\sqrt{2}}\,os = \frac{3}{2\sqrt{6}}\,os.$$

$$os = \frac{2\sqrt{2}}{\sqrt{3}}\,rds = \frac{2\sqrt{6}}{3}\,rds. \qquad (14.1)$$

14.2.3. SHORT-AXIS DISTANCE OF R.D. FACE

While we're at it, lets get $\overline{FG}$, the short-axis distance on the r.d. face. $\overline{FG}$ is visible in Figures 14.2, 14.3, 14.4, 14.5 and 14.8. We have already established $\overline{FX} = \frac{1}{2\sqrt{2}}\,os$. Substituting,

$$\overline{FX} = \frac{1}{2\sqrt{2}}\left(\frac{2\sqrt{2}}{\sqrt{3}}\,rds\right) = \frac{1}{\sqrt{3}}\,rds.$$

$$\text{Since } \overline{FG} = 2\left(\overline{FX}\right),$$

$$\overline{FG} = \frac{2}{\sqrt{3}}rds. \qquad (14.2)$$

14.2.4. DISTANCE FROM CENTROID TO SHORT AXIS POINTS OF THE R.D. FACE

It is important to establish the distance $\overline{OF} = \overline{OG}$, the distance from the r.d. centroid to the short axis points on the r.d. face. We have shown these in gold (Figures 14.5–14.8), indicating their length is equal to the length of the r.d. side. Is this true?

Triangle $\triangle OXF$ is right, by construction. In order to get $\overline{OF} = \overline{OG}$, we need $\overline{OX}$, which also happens to be the height of the r.d. pyramid.

The height of the pyramid can be determined by inspection. From the previous figures we see that $\overline{OX}$ goes from the centroid of the r.d. to the midpoint of the octahedron side. But this distance is exactly one-half the octahedron side, as we see clearly in Figure 14.11.

Figure 14.11 shows that the height of the r.d. pyramid ($\overline{OX}$) is one-half os, the side of the octahedron ($\overline{NI}, \overline{JM}$). So

$$h = \overline{OX} = \frac{1}{2}\,os, \text{ and } os = \frac{2\sqrt{2}}{\sqrt{3}}\,rds.$$

$$\text{Therefore } h = \frac{1}{2}\left(\frac{2\sqrt{2}}{\sqrt{3}}\right) = \frac{\sqrt{2}}{\sqrt{3}}\,rds.$$

Now we have the height of the r.d. pyramid in terms of the r.d. side.

$$\overline{XN} = \frac{1}{2}\,os = \frac{1}{2}\left(\frac{2\sqrt{2}}{\sqrt{3}}\,rds\right) = \frac{\sqrt{2}}{\sqrt{3}}\,rds.$$

The triangle $\triangle OXN$ is isosceles.

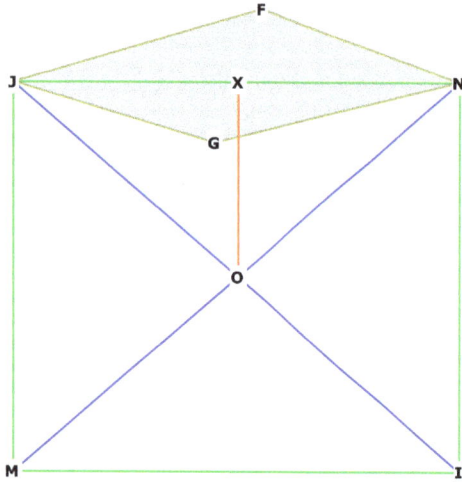

Figure 14.11: Showing the height of the r.d. pyramid $\overline{OX}$ is one=half os, the side of the octahedron

Now that we have calculated $\overline{OX}$, it remains to prove that $\overline{OF} = \overline{OG}$. When you build a model of the rhombic dodecahedron you can see this immediately, but lets show it mathematically as well. We will find the result first in terms of the side of the octahedron, then translate this with respect to the side of the r.d.

$$\overline{OF}^2 = \overline{OG}^2 = \overline{OX}^2 + \overline{XF}^2$$

$$= \frac{1}{4}\,os^2 + \frac{1}{8}\,os^2 = \frac{3}{8}\,os^2.$$

$$\overline{OF} = \overline{OG} = \frac{\sqrt{3}}{2\sqrt{2}}\,os,$$

which is precisely what we established above.

$$\overline{OF} = \overline{OG} = \frac{\sqrt{3}}{2\sqrt{2}}\left(\frac{2\sqrt{2}}{\sqrt{3}}\,rds\right) = rds. \qquad (14.3)$$

Therefore $\overline{OF} = \overline{OG}$ is equal in length to the side of the r.d.

14.2.5. THE RELATIONSHIP BETWEEN $\overline{OQ}$ AND $\overline{FQ}$

See Figure 14.9. $\overline{OQ}$ is the distance between the centroid and the mid-face of the internal octahedron. $\overline{FQ}$ is the distance from the mid-face of the internal octahedron to any r.d. vertex. What is the relationship between $\overline{OQ}$ and $\overline{FQ}$? From *Tetrahedron* and *Octahedron* we know that the distance from the centroid of the tetrahedron to the mid-face $= \frac{1}{2\sqrt{6}}\,os = \overline{FQ}$.

The distance from the centroid of the octahedron to the mid-face is

$$\frac{1}{\sqrt{6}}\,os = \overline{OQ}. \qquad (14.4)$$

Therefore $\overline{OQ} = 2\left(\overline{FQ}\right)$, and we conclude that the face of the octahedron is twice as far away from its centroid, as is the face of the tetrahedron from its centroid.

We also gain the important information that triangle $\triangle FOG$ is isosceles.

14.2.6. Rhombic Dodecahedron Side to Unit Sphere Containing Rhombic Dodecahedron

Now we can get to some important data, that is, the relationship between the side of the r.d. and the radius of the sphere that contains the rhombic dodecahedron. That sphere is, you recall, the sphere that surrounds and touches the vertices of the octahedron (N-ILJK-M in Figures 14.2 and 14.3). We will refer to this radius as r_{outer}.

The radius of this sphere is $\overline{ON} = \overline{OJ}$ in Figure 14.12 below. What is $\overline{ON} = \overline{OJ}$?

$\angle JON$ is right by construction.

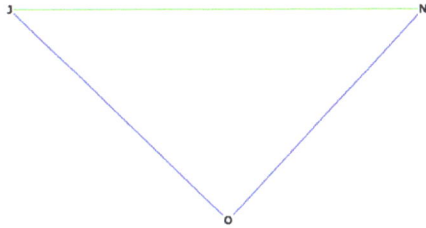

Figure 14.12: Radius of outer sphere of r.d.

We know that $\overline{JN}$ = side of the octahedron = os. $\overline{OJ} = \overline{ON} = r_{outer}$, so

$$2\left(r_{outer}^2\right) = os^2, \ r_{outer}^2 = \frac{os^2}{2}.$$

$$r_{outer} = \frac{1}{\sqrt{2}}os. \tag{14.5}$$

We found that $rds = \frac{\sqrt{3}}{2\sqrt{2}}os$, so we substitute for os and get,

$$r_{outer} = \frac{1}{\sqrt{2}}\left(\frac{2\sqrt{2}}{\sqrt{3}}rds\right) = \frac{2}{\sqrt{3}}rds. \tag{14.6}$$

Therefore

$$r_{outer} = \frac{2}{\sqrt{3}}rds, \ rds = \frac{\sqrt{3}}{2}r_{outer}. \tag{14.7}$$

Now we have expressed the side of the r.d. in terms of the radius of the sphere that contains the rhombic dodecahedron, realizing that this sphere touches only the 6 outer vertices of the r.d. that comprise an octahedron.

Notice that $\overline{FG}$, the short-axis distance across the face of the r.d., (calculated above), is also $\frac{2}{\sqrt{3}}rds$.

We also know from Figure 14.3 that the radius of the smaller sphere touching the 6 cube vertices ABCD-EFGH is equal to the side length of the r.d. We may write $rds = r_{inner}$ for this smaller sphere.

(Henceforth, we will write r for the radius, understanding that $r = r_{outer}$).

14.3. Resumption of the Volume Calculation

We were, at the end of the previous section, about to find the volume of the rhombic dodecahedron. To do this we need to recall the height of the pyramid $\overline{OX}$ (see Figure 14.11), and find the area of the r.d. face.

By looking at Figure 14.5 we perceive that the area of the r.d. face can be divided into two triangles, each one with base $\overline{NJ}$. Both triangles are congruent by side-side-side, and the height, ht, of each is $\overline{XF} = \overline{XG}$.

From above we know that this distance is $\frac{1}{2\sqrt{2}}\, os$.

$\overline{NJ}$ is the side of the octahedron, so $\overline{NJ} = os$.

Now we have

$$Area_{1\ \text{face triangle}} = \frac{1}{2}\,(\text{base})\,(\text{height}) = \frac{1}{2}\,(os)\left(\frac{1}{2\sqrt{2}}\,os\right) = \frac{1}{4\sqrt{2}}\,os^2.$$

$$Area_{1\ \text{face}} = 2\left(Area_{1\ \text{face triangle}}\right) = \frac{1}{2\sqrt{2}}\,os^2. \tag{14.8}$$

We want to get all of our data on the r.d. in terms of the r.d. itself, for consistency. Therefore we translate the area from the side of the octahedron, to the side of the r.d.

$$\begin{aligned}
Area_{1\ \text{face}} &= \frac{1}{2\sqrt{2}}\,os^2 = \frac{1}{2\sqrt{2}}\left(\frac{2\sqrt{2}}{\sqrt{3}}\right)^2 rds^2 \\
&= \frac{1}{2\sqrt{2}}\left(\frac{8}{3}\,rds^2\right) \\
&= \frac{4}{3\sqrt{2}}\,rds^2 = 0.942809042\,rds^2. \tag{14.9}
\end{aligned}$$

Now, finally, we have enough data to calculate the volume of 1 pyramid:

$$\begin{aligned}
Volume_{1\ \text{pyramid}} &= \frac{1}{3}\,(\text{area of base})\,(h) \\
&= \frac{1}{3}\left(\frac{4}{3\sqrt{2}}\,rds^2\right)\left(\frac{\sqrt{2}}{\sqrt{3}}\,rds\right) \\
&= \frac{4}{9\sqrt{3}}\,rds^3 \\
Volume_{\text{r.d.}} &= 12\left(Volume_{1\ \text{pyramid}}\right) \\
&= 12\left(\frac{4}{9\sqrt{3}}\,rds^3\right) \\
Volume_{\text{r.d.}} &= \frac{16}{3\sqrt{3}}\,rds^3 = 3.079201436\,rds^3. \tag{14.10}
\end{aligned}$$

14.4. Surface Area of Rhombic Dodecahedron

The surface area of the rhombic dodecahedron is just 12 faces times the area of 1 face:

$$12\left(\frac{4}{3\sqrt{2}}rds^2\right) = \frac{16}{\sqrt{2}}rds^2 = 8\sqrt{2}\,rds^2.$$

$$Surface\ area_{\text{rhombic dodecahedron}} = 8\sqrt{2}\,rds^2 = 11.3137085\,rds^2. \tag{14.11}$$

14.5. Central Angles of the Rhombic Dodecahedron

There are 3 central angles of the rhombic dodecahedron.

In Figure 14.13, they would be, for example, $\angle FOG$, $\angle FON$, $\angle JON$.

Let's start with $\angle FOG$.

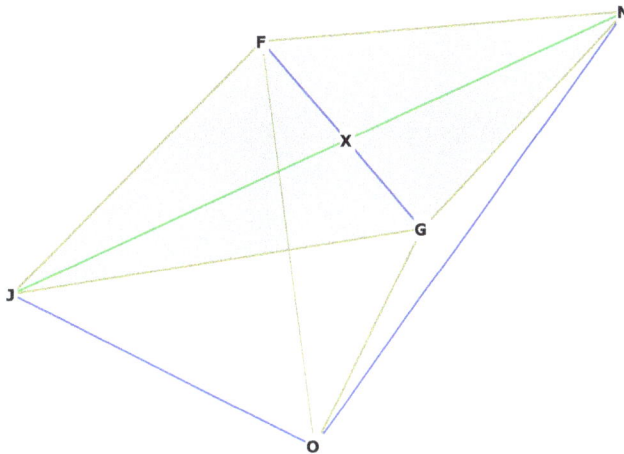

Figure 14.13: Central angles of rhombic dodecahedron (Figure 14.5, repeated)

$\angle JON$ is right. $\angle OXF$, $\angle OXG$ are right.

To find $\angle FOG$, find $\angle FOX$. Triangle $\triangle FOX$ is right by construction, therefore,

$$\sin(\angle FOX) = \frac{\overline{FX}}{\overline{FO}} = \frac{\frac{1}{\sqrt{3}}}{1} = \frac{1}{\sqrt{3}},$$

$$\angle FOX = \arcsin\left(\frac{1}{\sqrt{3}}\right) = 35.26438968°,$$

$$\angle FOG = 2\,(\angle FOX) = 70.52877936°. \tag{14.12}$$

This is precisely the central angle of the cube! Understandably so, for we already know that the two vertices F and G are two of the vertices of a cube (see Figure 14.2).

It is also the dihedral angle of the tetrahedron.

Because the triangle $\triangle FOG$ is isosceles,and using the property that the sum of angles of a triangle equal $180°$,

$$\angle OFG = \angle OGF = \angle FON = \frac{(180° - 70.52877936°)}{2}.$$

Therefore,

$$\angle FON = 54.7356103°, \text{and so does } \angle FNO. \tag{14.13}$$

From Figure 14.11 we can see immediately that $\angle JON$ is right, it being the central angle of the square JMIN.

$$\angle JON = 90°. \tag{14.14}$$

14.6. Surface Angles of the Diamond Faces of the Rhombic Dodecahedron

We know that the short-axis distance across the r.d. face, $\overline{FG}$ in Figure 14.5, is $\frac{2}{\sqrt{3}}\,rds$.

We know that $\overline{FN} = \overline{NG}$ = the side of the r.d. $= rds$.

Triangle $\triangle FXN$ is right by construction.

$\triangle FNX$ is one-half the face angle $\angle FNG$. $\overline{FX}$ is one-half $\overline{FG}$.

So we can write:

$$\sin(\angle FNX) = \frac{\overline{FX}}{\overline{FN}} = \frac{1}{\sqrt{3}}.$$

$\angle FNX = 35.26438968°$, so

$$\angle FNG = 70.52877936°. \tag{14.15}$$

$\angle FNG$ and its counterpart $\angle FJG$ are the smaller of the two face angles of the r.d. Now let's get $\angle JFN$, the larger face angle.

Once again we use triangle $\triangle FXN$ and work with $\angle XFN$, which is one-half the desired angle, $\angle JFN$.

$\angle XFN$ is just $90 - \angle FNX$, using the property that the sum of all angles in a triangle is $180°$.

$\angle XFN = 54.73561032°$, so

$$\angle JFN = 109.4712206°. \tag{14.16}$$

This angle, $109.4712206°$, is the central angle of the tetrahedron and the dihedral angle of the octahedron. It also is the angle you see when you stand the rhombic dodecahedron on one of its eight octahedral vertices and look down from above at two intersecting r.d. sides. This polyhedron is an all-space filler, meaning that it can be joined with itself, like the cube, to fill any volume without any space left over.

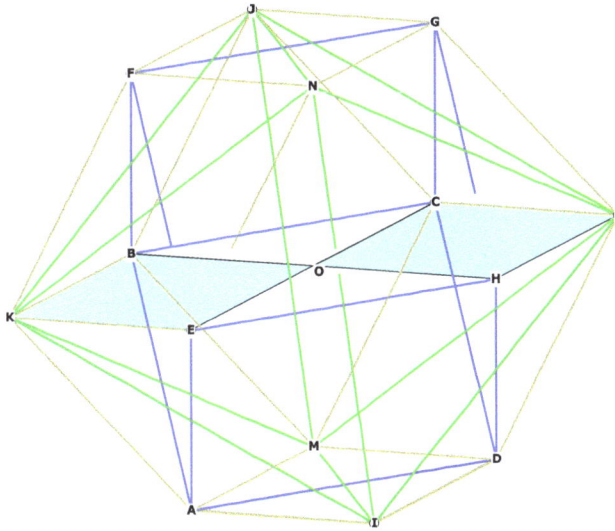

Figure 14.14: Two of the 12 internal rhombi of the rhombic dodecahedron

14.7. Internal Rhombi of the Rhombic Dodecahedron

If you examine the r.d. from the inside out with the zometool, you will see it is composed internally of the same rhombi as the faces.

To identify these internal rhombuses, lay the rhombic dodecahedron flat on one of its faces. Find three r.d. points and the centroid to see the rhombus:

There are 12 internal rhombi, and 12 external rhombi, each of them identical.

Note that there is another angle on the exterior of the rhombic dodecahedron, and that is the angle that one face plane makes with another face plane as it goes over top of the octahedron within the r.d. This angle is 90°. This can be seen by looking at the long axis diagonals of the octahedron through the rhombi, and following the faces that form around the side of the octahedron. In Figure 14.5, this can be seen with the faces NFJG and JBMC. Notice the lines $\overline{NJ}$ and $\overline{JM}$, which are part of the octahedral square, and which bisect the faces along their long axis. $\angle NJM$ is a right angle, as is $\angle NLM$ and $\angle NKM$. Observe also $\angle KNL$. Again, this property helps the rhombic dodecahedron to be an all-space filler.

14.8. The Dihedral Angle of the Rhombic Dodecahedron

When calculating dihedral angles, it is vital to ensure that the angle chosen is an accurate representation of the intersection of the two planes. In the rhombic dodecahedron (r.d.), this is a little tricky.

The dihedral angle of the r.d. is the angle $\angle EXG$ in Figures 14.15 and 14.16. This angle is a geometrically accurate description of the intersection of any two planes of the r.d., in this case, the two

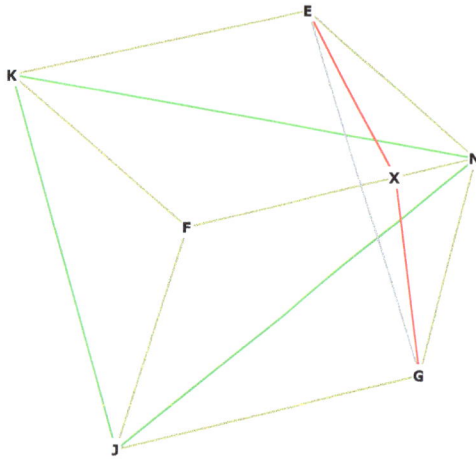

Figure 14.15: The rhombic dodecahedron dihedral angle

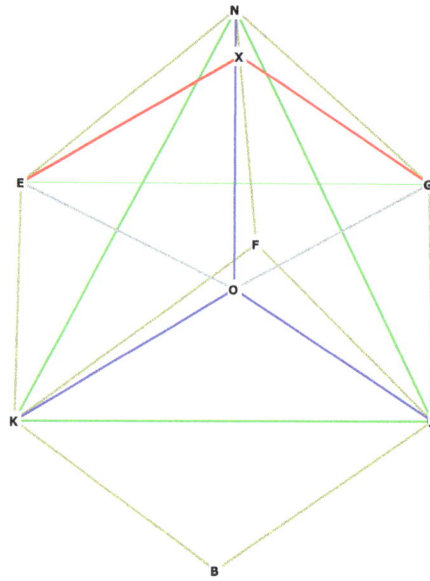

Figure 14.16: Another view of the r.d. dihedral angle

planes KENF and JGNF. The lines $\overline{EX}$ and $\overline{GX}$ are constructed such that $\angle NXG$ and $\angle NXE$ are right. $\angle EXG$ is a "roof" over the two planes which exactly describes the dihedral angle.

We will be working with triangle $\triangle EXG$.

It's easier to see this if you construct a model of the rhombic dodecahedron, and lay it on one of its edges.

I have shown $\overline{EG}$ in green, indicating it is equal to the side of the octahedron (referred to as os). We have to show this.

In Figure 14.16, O is the centroid. NKJ is one of the faces of the octahedron. $\overline{ON}, \overline{OK},$ and $\overline{OJ}$ are radii of the outer sphere that touch the 6 octahedral vertices. $\overline{OE}$ and $\overline{OG}$ touch the short axis vertices of the r.d.

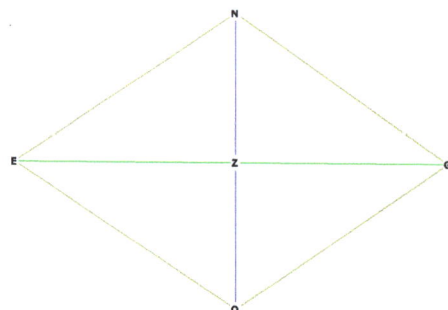

Figure 14.17 shows another view, showing how $\overline{EG}$ is found. NEOG is one of 12 internal rhombi of the rhombic dodecahedron.

Figure 14.17: Rhombic dodecahedron internal rhombus

$\overline{OE}$ and $\overline{OG}$ are the sides of the rhombic dodecahedron, or rds. We have previously shown that the triangles $\triangle OEN$ and $\triangle OGN$ are isosceles and congruent (the vertices may have different names, but the triangles are the same!) $\angle NZE$ is right.

To obtain $\overline{EG}$, simply re-do the calculation we did previously,

$$\overline{EZ}^2 = \overline{EN}^2 - \overline{NZ}^2 = rds^2 - \frac{1}{3}rds^2 = \frac{2}{3}rds^2,$$

$$\overline{EZ} = \frac{\sqrt{2}}{\sqrt{3}}rds.$$

$$\overline{EG} = 2\left(\overline{EZ}\right),$$

$$\overline{EG} = \frac{2\sqrt{2}}{\sqrt{3}}rds, \tag{14.17}$$

which has been obtained previously as os.

Therefore $\overline{EG}$ is identical in length to the side of the octahedron.

This internal rhombus, NEOG, is identical to the rhombus of the r.d. face. The r.d. is composed, internally and externally, of identical rhombi, which gives it the quality of an all-space filler.

We have established $\overline{EG}$ as the side of the octahedron. Now it remains to determine $\overline{GX}$ and $\overline{XB}$.

Look at Figure 14.15 again. The triangle $\triangle ENX$ is right by construction, and we know the angle $\angle ENX$, it being the long-axis angle of the r.d. face, as $2\left[\arcsin\left(\frac{1}{\sqrt{3}}\right)\right]$, or $70.52877936°$.

We also know $\overline{EN}$ = side of the r.d, or rds.

So we write $\sin(\angle ENX) = \frac{\overline{EX}}{\overline{EN}} = \frac{\overline{EX}}{rds}$.

$$\overline{EX} = rds \cdot \sin\left(70.52877936°\right) = \frac{2\sqrt{2}}{3}rds = 0.942809042\,rds.$$

(Note: $\overline{EX} = \frac{1}{\sqrt{3}}os$).

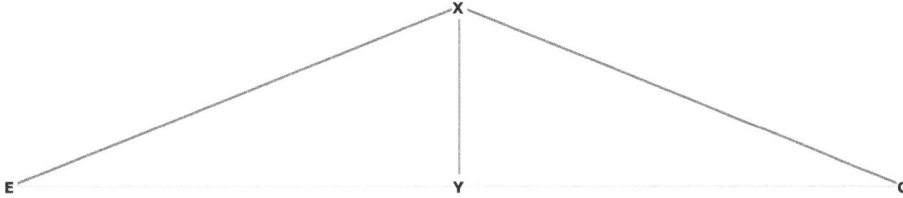

Figure 14.18: Showing the dihedral angle in simplified fashion

Now, $\overline{EX}$ and $\overline{EY}$ are known. We need to get $\angle EXY$ to get $\angle EXG$, the dihedral angle.

We know $\overline{EG}$ is the side of the octahedron, or os, and it is $\frac{2\sqrt{2}}{\sqrt{3}}rds$, and so $\overline{EY}$ is one-half of that.

$$\sin(\angle EXY) = \frac{\overline{EY}}{\overline{EX}} = \frac{\frac{\sqrt{2}}{\sqrt{3}}rds}{\frac{2\sqrt{2}}{3}rds} = \frac{3}{2\sqrt{3}} = \frac{\sqrt{3}}{2}.$$

$$\overline{EXY} = \arcsin\left(\frac{\sqrt{3}}{2}\right) = 60°.$$

Therefore

$$\angle EXG, \text{the dihedral angle,} = 2\left(\angle EXY\right) = 120°. \tag{14.18}$$

14.9. Centroid Distances

Now let's complete the analysis of the rhombic dodecahedron by finding or collecting the distances from the centroid to a short-axis vertex, a long-axis vertex, mid-edge and mid-face. We already have all of this information, except for the distance from the centroid to a mid-edge:

$$\text{distance from centroid to long-axis r.d. vertex (vertex of the octahedron)} = \frac{2}{\sqrt{3}}\,rds. \quad (14.19)$$

$$\text{distance from centroid to short-axis r.d. vertex (vertex of the cube)} = rds. \quad (14.20)$$

$$\text{distance from centroid to r.d. mid-face (mid-point of octahedron side)} = \frac{\sqrt{2}}{\sqrt{3}}\,rds. \quad (14.21)$$

To find the centroid to r.d. mid-edge distance requires a little work.

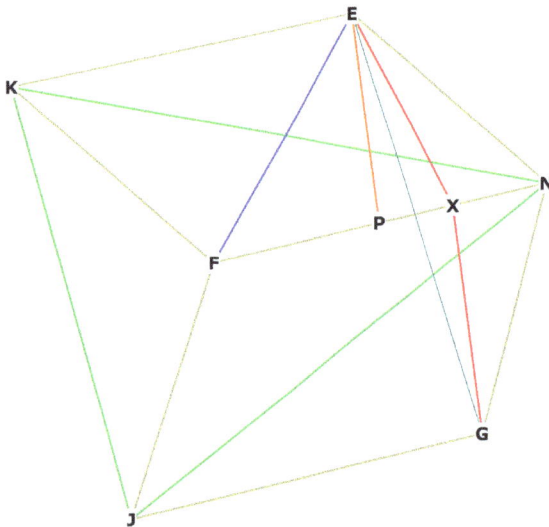

Figure 14.19: Centroid to mid-edge distance $\overline{EP}$

We have already remarked upon one of the properties of the r.d., that it is composed of 12 internal rhombi identical to the external faces. Those rhombi are divided in half (here, at $\overline{EF}$) by the same distance as from the centroid to any of the vertices of the octahedron contained within the r.d. Therefore the r.d. is composed of 24 exterior congruent triangles (here, $\angle EFN$ and $\angle EKF$), and 24 interior congruent triangles congruent to the exterior ones!

So triangle $\triangle EFN$ may substitute for any triangle in the interior of the r.d., and more to the point, the vertex E or F may substitute for the centroid of the r.d., at O. In order to see this clearly, you have to build a 3D model of the rhombic dodecahedron, and I encourage you to do this.

From *Rhombic Dodecahedron Dihedral Angle* (p. 123), we know that $\overline{EX} = \frac{2\sqrt{2}}{3}\,rds$.

We know that $\angle EXN$ is right by construction, and that $\overline{EN} = rds$. We also know that P is at mid-edge on $\overline{FN}$, the side of the r.d., therefore $\overline{NP} = \frac{1}{2}\,rds$.

We can find $\overline{XN}$, and so $\overline{XP}$, by writing $\overline{XP} = \overline{NP} - \overline{XN}$.

We need to find $\overline{XN}$.

$$\overline{XN}^2 = \overline{EN}^2 - \overline{EX}^2 = 1\,rds^2 - \frac{8}{9}\,rds^2 = \frac{1}{9}\,rds^2.$$

$$\overline{XN} = \frac{1}{3}\,rds.$$

$$\overline{XP} = \frac{1}{2}\,rds - \frac{1}{3}\,rds = \frac{1}{6}\,rds.$$

Now we can find $\overline{EP}$, the distance from centroid to mid-edge.

$$\overline{EP}^2 = \overline{EX}^2 + \overline{XP}^2 = \frac{8}{9}\,rds^2 + \frac{1}{36}\,rds^2 = \frac{33}{36}\,rds^2 = \frac{11}{12}\,rds^2.$$

$$\overline{EP} = \frac{\sqrt{11}}{2\sqrt{3}}\,rds = 0.957427108\,rds. \tag{14.22}$$

14.10. Rhombic Dodecahedron Reference Tables

Table 14.1: Volume and Surface Area

Volume (edge)	Volume in Unit Sphere	Surface Area (edge)	Surface Area in Unit Sphere
$3.079201436\ s^3$	distance to outer sphere touching 6 octahedral vertices	$11.3137085\ s^2$	distance to outer sphere touching 6 octahedral vertices
	$2.0\ r^3$		$8.485281375\ r^2$

Table 14.2: Angles

Central Angles			Dihedral Angle	Surface Angles	
long axis	short axis	adjacent vertex	$120°$	long axis	short axis
$90°$	$70.52877936°$	$54.7356103°$		$109.4712206°$	$70.52877936°$

Table 14.3: Centroid Distances

Centroid To Vertex	Centroid To Mid-edge	Centroid To Mid-face
$1.0\ r$	$0.829156198\ r$	$0.707106781\ r$
$1.154700538\ s$	$0.957427108\ s$	$0.816496581\ s$

Table 14.4: Side to Radius

Side / radius	
to 8 cube vertices	to 8 octahedral vertices
1.0	0.866025404

Chapter 15

The Icosa-Docedahedron

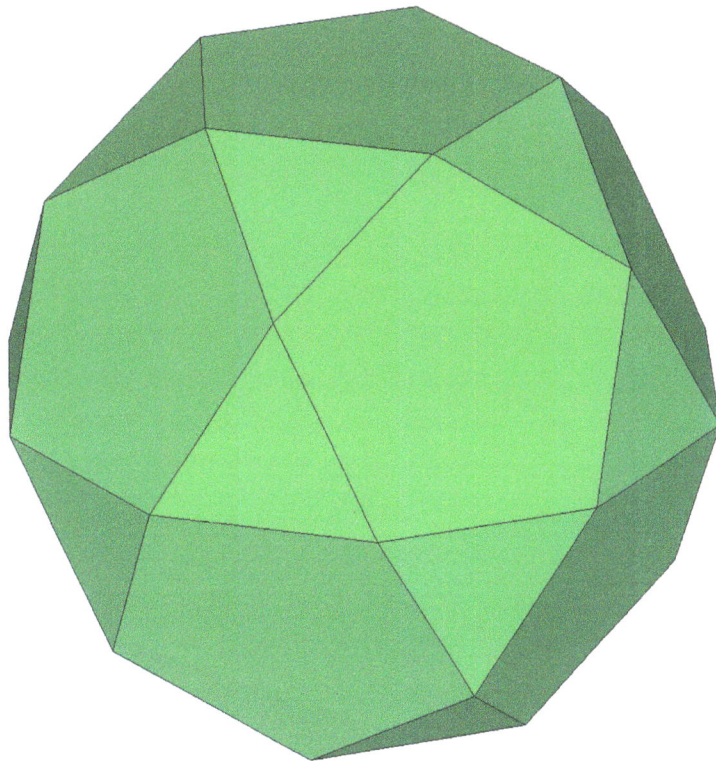

Figure 15.1: The Icosa-dodecahedron

15.1. Introduction

This polyhedron is the dual of the rhombic triacontahedron.

It has 30 vertices, 32 faces, and 60 edges.

Twenty of the faces are equilateral triangles. Twelve of the faces are pentagons.

129

It is probably called the icosa-dodecahedron because in the middle of every pentagonal face is the vertex of an icosahedron, and in the middle of every triangular face is the vertex of a dodecahedron.

This polyhedron is composed of 6 "great circle" decagons which traverse the outside of the polyhedron, sharing the 30 vertices and accounting for the 60 edges (see Figure 15.2, each decagon is marked with a different color).

Actually, the rhombic triacontahedron should be called the icosa dodecahedron, because one icosahedron and one dodecahedron precisely describe its 32 vertices.

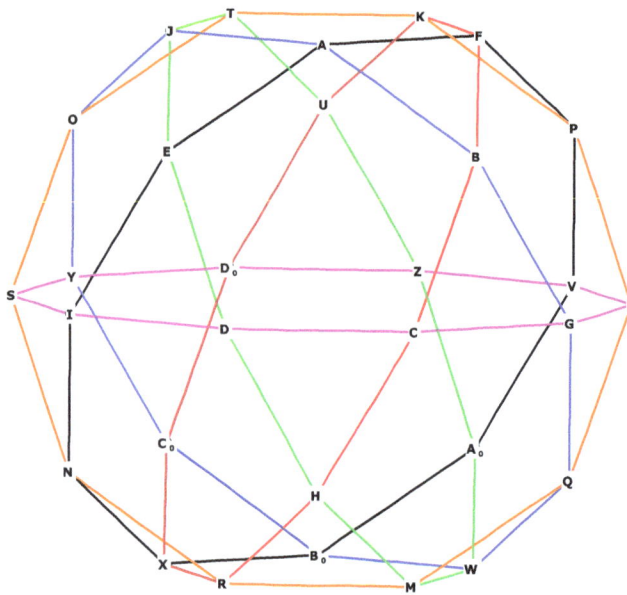

Figure 15.2: The 6 interlocking "great decagons" that compose the i.d.

The icosadodecahedron (hereinafter referred to as i.d.) has internal planes within it, as we have seen in other polyhedra. The i.d. has 12 internal pentagons, two of which are highlighted, one for each pentagonal face. And of course, the 6 "great circle" decagons:

Fascinatingly enough, the i.d. also has 5 hexagonal planes! One of these is highlighted in Figure 15.4 below, the plane $LBESC_0A_0$.

Notice that the sides of the hexagonal planes are, just like the pentagonal planes, all diagonals of the pentagonal faces. How can this be? How can a pentagon and a hexagon both have sides of the same length? We'll see later on how this happens.

The i.d. is composed of equilateral triangles internally and externally. The internal equilateral triangles form from the centroid and any two diagonal vertices of the pentagonal faces.

In Figure 15.5 on page 132, all of these internal triangles (for example, triangle BG_0E) are equilateral triangles, just like the 20 smaller equilateral triangles of the faces (for example, AEJ). The i.d. has the property that the central angles of the diagonals of the pentagonal faces (as BG_0E) are 60 degrees. Note also that the sides of these internal equilateral triangles (as in $\overline{BE}$) are diagonals of the pentagonal faces (as in ABCDE) of the icosa-dodecahedron.

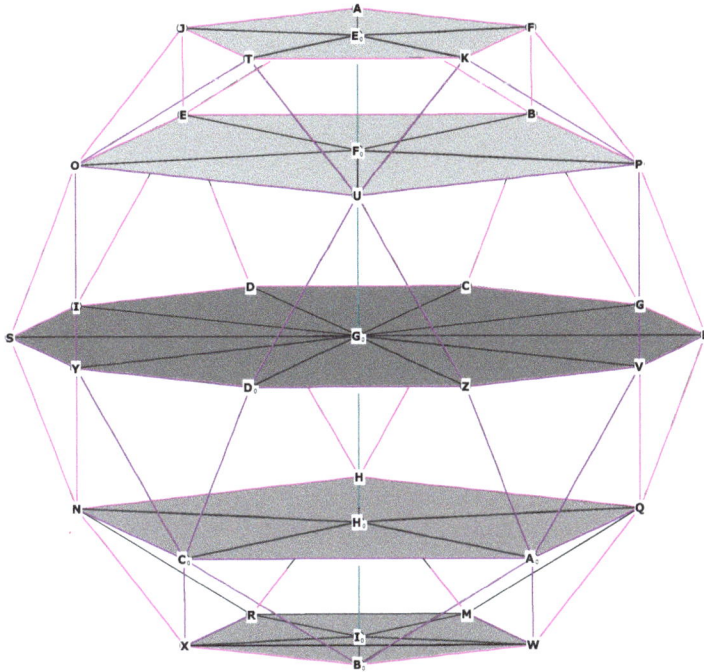

Figure 15.3: Four of the internal pentagons and 1 internal decagon

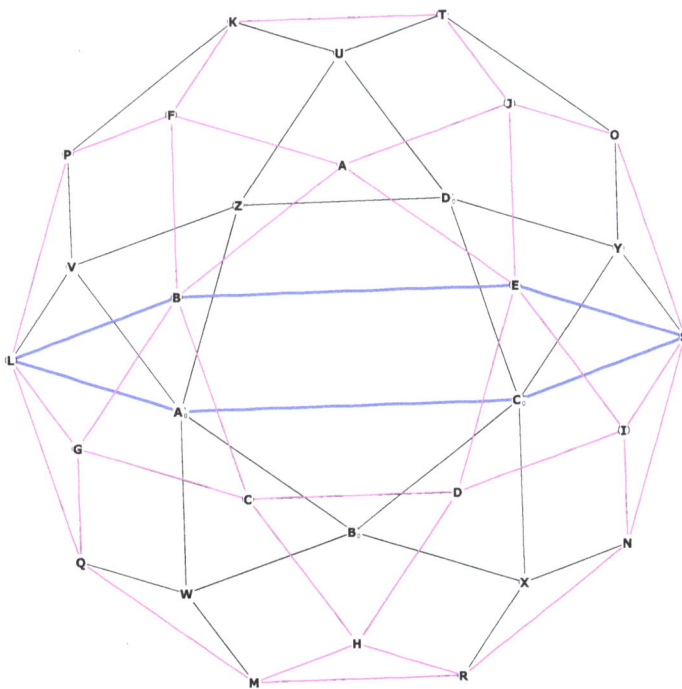

Figure 15.4: One of the 5 internal hexagonal planes of the icosa-dodecahedron

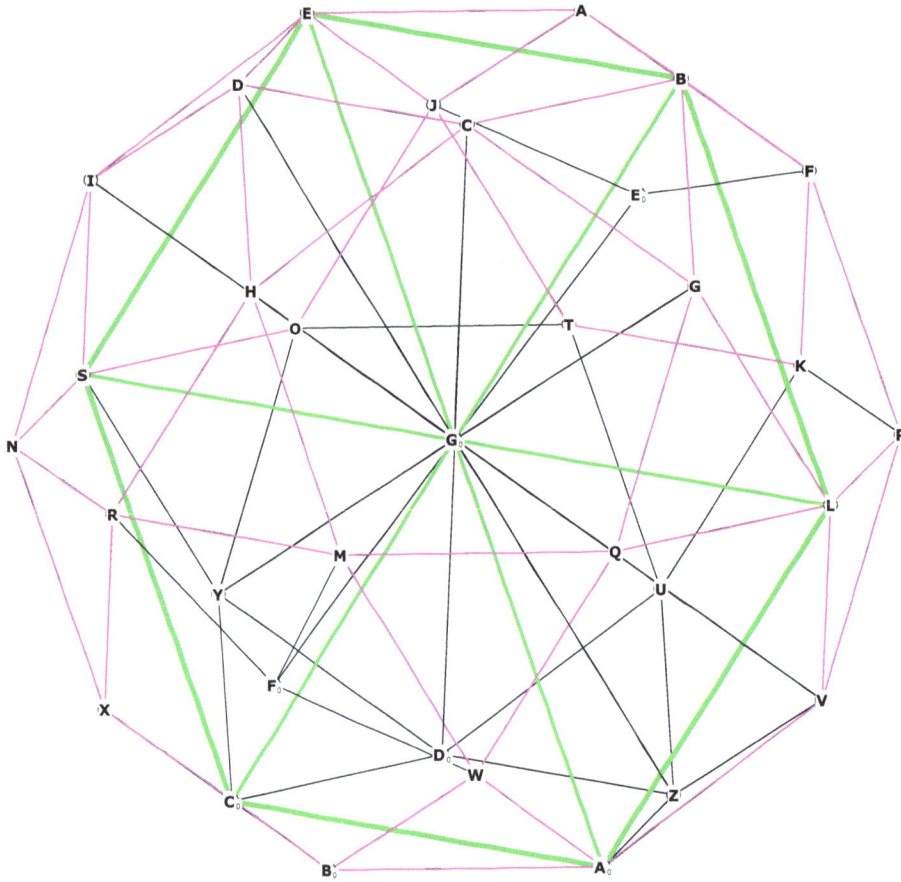

Figure 15.5: Another view of Figure 15.4, showing the internal hexagonal plane composed of 6 equilateral triangles

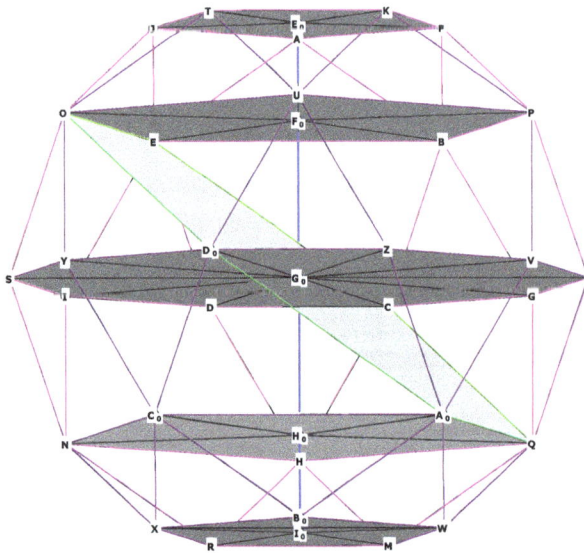

Figure 15.6: Showing one of the hexagonal planes in relation to two of the pentagonal planes and one of the dodecahedron planes.

Examining Figure 15.6, it is now clear why the hexagonal plane has the same edge length as the pentagonal planes: the hexagonal plane uses different diagonals of the pentagonal faces, and so is angled relative to the pentagonal planes.

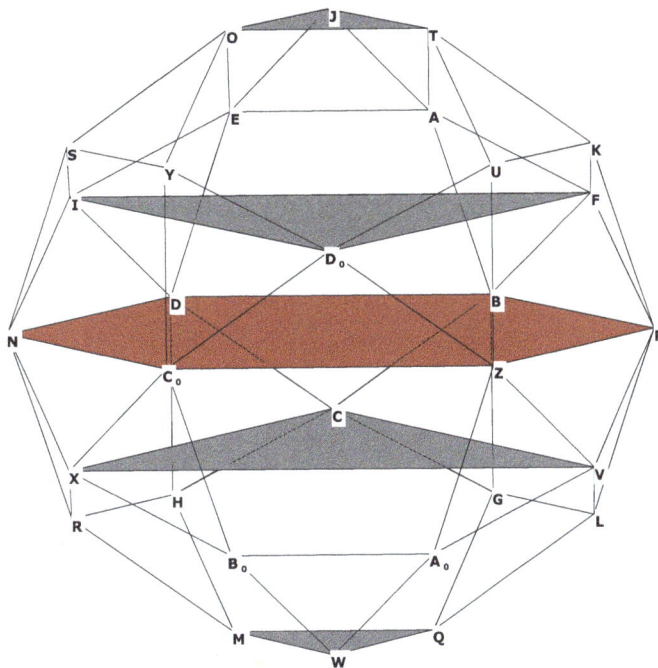

Figure 15.7: Showing that the internal hexagonal plane is parallel to the top and bottom triangular faces

15.2. Volume of the Icosadodecahedron

We will use the pyramid method again. There are 20 triangular pyramids and 12 pentagonal pyramids.

Although the icosa-dodecahedron is a semi-regular polyhedron, all of its vertices touch on the same sphere. So the radius of the circumsphere surrounding the i.d. is the same for each vertex. That makes our calculations a little easier.

In order to make life simpler, it would be nice to know the relationship between the side and the radius of the i.d. I will present two ways of getting this: the absurdly simple way, and the 'brute force' method.

The simple method is to recognize that the central angle of the i.d. is 36°, and that every vertex on the i.d. is part of a decagon. Therefore any two adjacent vertices combined with the centroid forms a 36–72–72 isosceles triangle. We recognize this immediately as a Golden Triangle. The relationship of the short side to any of the long sides of such a triangle is therefore 1 to Φ. And so the relationship of the radius to any edge of the i.d. is

$$r = \Phi \cdot ids, \tag{15.1}$$

where ids is the icosadodecahedron side.

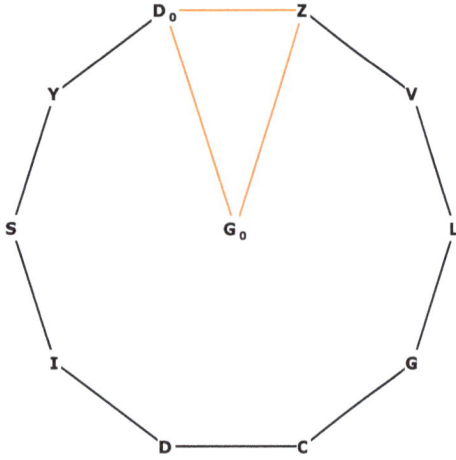

Figure 15.8: $\overline{D_0 G_0} = \overline{G_0 Z}$ = radius of circumsphere, and the golden triangle Go-Z-Do

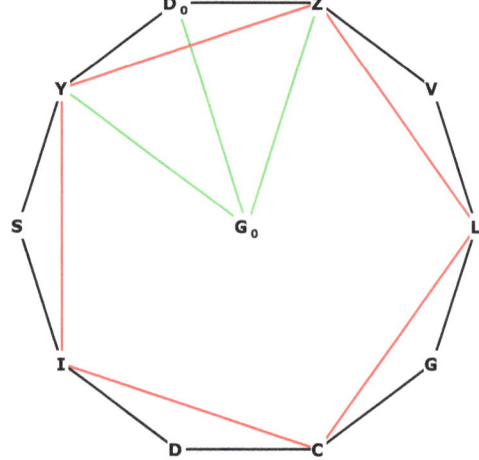

Figure 15.9: Deriving the relationship between radius ($\overline{G_0 Z}$) and side ($\overline{ZD_0}$) of the icosa dodecahedron

We can calculate this relationship as follows:

$\overline{ZY}$ is the side of the pentagon. G_0 is also the center of the pentagon.

$\overline{ZT}$ is just one half the side of the pentagon, because $\overline{G_0 D_0}$ is an angle bisector of $\overline{ZG_0 Y}$.

We know the distance $\overline{G_0 Z}$, from *Construction of the Pentagon Part 2*, as $\left(\frac{\Phi}{\sqrt{\Phi^2 + 1}} \right)$ side of pentagon.

$\overline{G_0 Z}$ is also the radius of the enclosing sphere around the i.d.

We also know from this that $\overline{G_0 T} = \frac{\Phi^2}{2\sqrt{\Phi^2 + 1}}$ (side of pentagon).

Therefore,

$$\overline{TD_0} = \frac{\Phi}{\sqrt{\Phi^2 + 1}} - \frac{\Phi^2}{2\sqrt{\Phi^2 + 1}} = \frac{1}{2\Phi\sqrt{\Phi^2 + 1}} \text{ (side of pentagon)}.$$

15.2.1. The Relationship between the Side of the Pentagon and the Side of the Decagon

Let's find $\overline{ZD_0}$, the side of the decagon, in terms of the side of the pentagon.

$$\overline{ZD_0}^2 = \overline{TD_0}^2 + \overline{TZ}^2$$

$$= \frac{1}{4\Phi^2 (\Phi^2 + 1)} + \frac{1}{4} = \frac{1 + \Phi^2 (\Phi^2 + 1)}{4\Phi^2 (\Phi^2 + 1)} = \frac{4\Phi^2}{4\Phi^2 (\Phi^2 + 1)}$$

$$= \frac{1}{\Phi^2 + 1} \cdot \text{(side of pentagon)}. \tag{15.2}$$

$\overline{ZD_0} = \frac{1}{\sqrt{\Phi^2 + 1}} \cdot$ (side of pentagon). But $r = \frac{\Phi}{\sqrt{\Phi^2 + 1}} \cdot$ (side of pentagon), so side of pentagon $= \frac{\sqrt{\Phi^2 + 1}}{\Phi} r$. Therefore, $\overline{ZD_0}$, the side of the decagon and the side of the i.d., is

$$\frac{1}{\sqrt{\Phi^2 + 1}} \left(\frac{\sqrt{\Phi^2 + 1}}{\Phi} \right) = \frac{1}{\Phi} r.$$

$$r = \Phi\left(ids\right), \; ids = \frac{1}{\Phi}\, r. \tag{15.3}$$

This calculation confirms the result above.

15.3. Resumption of the Volume Calculation

First we'll calculate the volume of a pentagonal pyramid. Refer to Figure 15.10.

From section 8.8 *Area of the Pentagon* we know that the area of a pentagon is $\frac{5\Phi^2}{4\sqrt{\Phi^2+1}}ids^2$, and the distance mid-face to any vertex of a pentagon, $\overline{H_0F}$, is $\frac{\Phi}{\sqrt{\Phi^2+1}}ids$.

$\overline{G_0F} = \Phi \cdot ids$. $\overline{G_0F}$ is just the radius of the enclosing sphere. To find the height of the pyramid, take right triangle $\triangle G_0 H_0 F$.

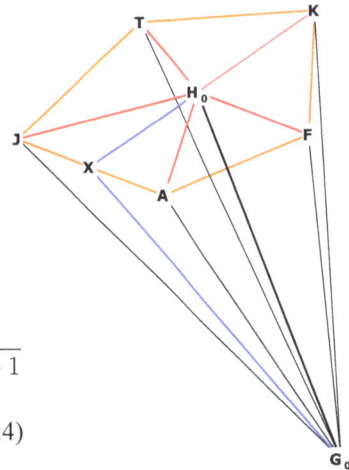
Figure 15.10: One of the 12 pentagonal pyramids of the i.d.

$$h^2 = \overline{G_0 H_0}^2 = \overline{G_0 F}^2 - \overline{H_0 F}^2$$

$$h^2 = \Phi^2 - \frac{\Phi^2}{\Phi^2+1} = \frac{\Phi^4 + \Phi^2 - \Phi^2}{\Phi^2+1} = \frac{\Phi^4}{\Phi^2+1}$$

$$h = \frac{\Phi^2}{\sqrt{\Phi^2+1}}\, ids. \tag{15.4}$$

$$\left(\text{Note: } \frac{\overline{G_0 H_0}}{\overline{H_0 F}} = \frac{\frac{\Phi^2}{\sqrt{\Phi^2+1}}\, ids}{\frac{\Phi}{\sqrt{\Phi^2+1}}\, ids} = \Phi \right).$$

Figure 15.11: A triangular pyramid of the icosadodecahedron

$$Volume_{1 \text{ pent pyramid}} = \frac{1}{3}\,(\text{area of base})\,(\text{height})$$

$$= \frac{1}{3}\left(\frac{5\Phi^2}{4\sqrt{\Phi^2+1}}\, ids^2\right)\left(\frac{\Phi^2}{\sqrt{\Phi^2+1}}\, ids\right)$$

$$= \frac{5\Phi^4}{12(\Phi^2+1)}\, ids^3$$

$$Volume_{12 \text{ pyramids}} = \frac{5\Phi^4}{\Phi^2+1}\, ids^3. \tag{15.5}$$

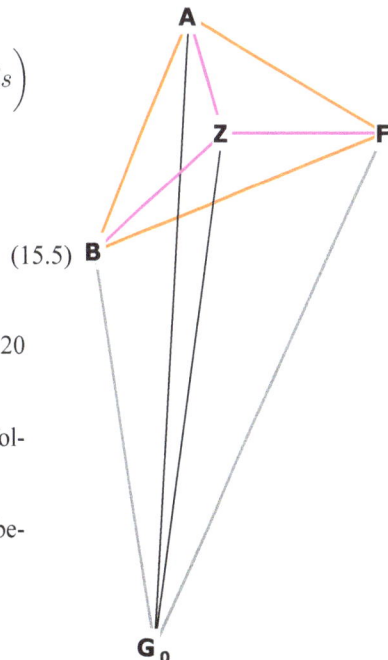

That takes care of the pentagonal pyramids. There are still 20 triangular pyramids.

Each triangular face area $= \frac{\sqrt{3}}{4}\, ids^2$, and the height is as follows:

Triangle $\triangle G_0 Z F$ is right by construction. $\overline{G_0 F} = \Phi\, ids$, being the radius of the enclosing sphere (circumsphere).

$\overline{ZF}$ is, from *Equilateral Triangle*, $\frac{1}{\sqrt{3}}\, ids$.

Therefore

$$h^2 = \overline{G_0 Z}^2 = \overline{G_0 F}^2 - \overline{ZF}^2 = \Phi^2 - \frac{1}{3} = \frac{3\Phi^2 - 1}{3} = \frac{\Phi^4}{3}.$$

$$h = \frac{\Phi^2}{\sqrt{3}}\, ids = 1.511522629\, ids. \tag{15.6}$$

So

$$Volume_{1\text{ triangular pyramid}} = \frac{1}{3}\,(\text{area of base})\,(\text{pyramid height})$$

$$= \frac{1}{3}\left(\frac{\sqrt{3}}{4}\, ids^2\right)\left(\frac{\Phi^2}{\sqrt{3}}\, ids\right)$$

$$= \frac{\Phi^2}{12}\, ids^3. \tag{15.7}$$

$$Volume_{20\text{ triangular pyramids}} = \frac{5\Phi^2}{3}\, ids^3. \tag{15.8}$$

$$Volume_{\text{total}} = Volume_{12\text{ pyramids}} + Volume_{20\text{ pyramids}}$$

$$= \frac{5\Phi^4}{\Phi^2 + 1}\, ids^3 + \frac{5\Phi^2}{3}\, ids^3$$

$$= \frac{15\Phi^4 + 5\Phi^2\left(\Phi^2 + 1\right)}{3\left(\Phi^2 + 1\right)}$$

$$= \frac{20\Phi^4 + 5\Phi^2}{3\left(\Phi^2 + 1\right)}$$

$$= \frac{5\Phi^2(4\Phi^2 + 1)}{3(\Phi^2 + 1)}\, ids^3$$

$$= 13.83552595\, ids^3 \tag{15.9}$$

This figure is slightly larger than the volume of the rhombic triacontahedron. This larger figure is easily explained by the fact that both the outer radial distance of the r.t., and the radius of the i.d., are $\Phi\,(side)$. Since ALL vertices of the icosadodecahedron lie on this sphere, and only 12 of the vertices of the rhombic triacontahedron do, the volume of the icosadodecahedron is naturally larger.

15.4. Surface Area of the Icosadodecahedron

The surface area of the i.d. is $12 \cdot$ area of pentagonal face $+20 \cdot$ area of triangular face.

$$Surface\,area = 12\left(\frac{5\Phi^2}{4\sqrt{\Phi^2 + 1}}\, ids^2\right) + 20\left(\frac{\sqrt{3}}{4}\right) ids^2$$

$$= \frac{15\Phi^2}{\sqrt{\Phi^2 + 1}}\, ids^2 + 5\sqrt{3}\, ids^2$$

$$= 29.30598285\, ids^2. \tag{15.10}$$

15.5. Central Angle of the Icosadodecahedron

Because all of the vertices of the i.d lie along the 6 "great circle" decagons, the interior angles are all $360°/10 = 36°$. Notice that each of the central angles to adjacent vertices forms a Golden Triangle with angles $36°$, $72°$, and $72°$.

We saw earlier how the central angle between two diagonal vertices on any of the pentagonal faces, is $60°$ and how this forms a hexagon, and thus large internal equilateral triangles (see Figure 15.6) and triangles $\triangle BG_0E$, $\triangle LBG_0$, $\triangle G_0ES$, $\triangle LA_0G_0$, $\triangle A_0G_0C_0$, and $\triangle G_0C_0S$).

We were curious as to how the side of the hexagonal plane could be the same length as the side of the pentagonal plane. If you look at Figure 15.6 you can see that the sides of both of these planes are a diagonal of any of the pentagonal faces. It is clear then, that like the dodecahedron and the icosahedron, the icosadodecahedron is based on the pentagon. Because the length of the radius (distance from centroid to any vertex) is $\Phi \cdot$ (side of i.d.), and the diagonal of any pentagon is also $\Phi \cdot$ (side of i.d.), the hexagonal plane is formed.

15.6. Surface Angles of the Icosadodecahedron

There are two surface angles: The angle forming the triangular faces, equal to $60°$, and the angle forming the pentagonal faces, equal to $108°$.

15.7. Dihedral Angle of the Icosadodeccahedron

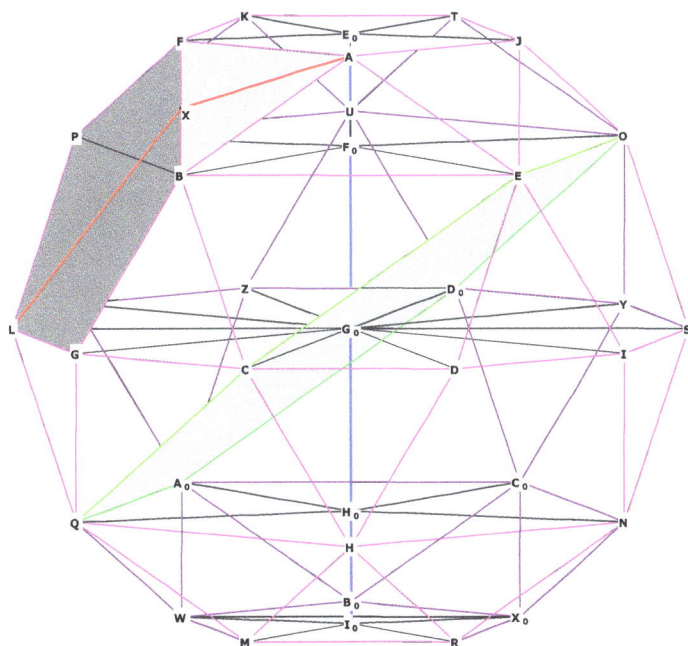

Figure 15.12: The dihedral angle of the icosa-dodecahedron – 3D

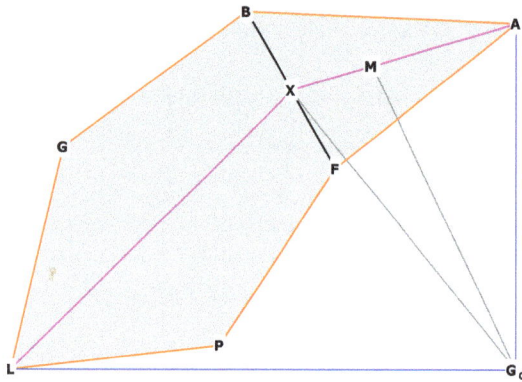

Figure 15.13: The dihedral angle of the icosa-dodecahedron – 2D

Consider Figure 15.13 and Figure 15.12. We see that the dihedral angle is $\angle LXA$, being the intersection of the two planes BGLPF and FBA. I have drawn the line $\overline{LX\text{-}XA}$ which goes directly through the middle of both planes and forms a "roof" over them.

If you build an icosa-dodecahedron you will see immediately that the angle from A to the centroid G_0 and back to L, or $\angle LOA$, is right.

$\overline{LO} = \overline{AO} = $ radius of enclosing sphere $= \Phi \cdot ids$.

Our plan of attack for getting $\angle LXA$ will be as follows:

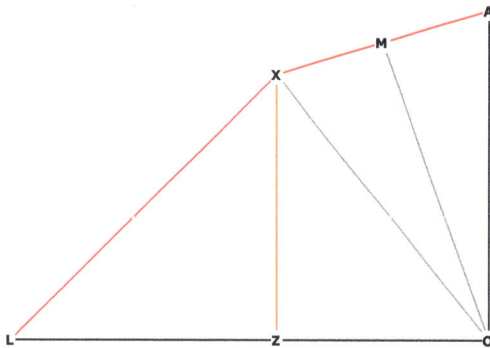

Figure 15.14: Dihedral angle simplified

Note that $\overline{ZX}$ is drawn parallel to $\overline{OA}$, and thus also perpendicular to $\overline{LO}$.

Therefore $\triangle XZO$ and $\triangle XZA$ are right.

$\overline{AM}$ and $\overline{VMX}$ are the distances, respectively, from the vertex of an equilateral triangle to the midface, and from the mid-face to a mid-edge.

From *Equilateral Triangle*, we know these to be

$$\frac{1}{\sqrt{3}} \, ids \ \text{ and } \ \frac{1}{2\sqrt{3}} \, ids.$$

$\overline{XL}$ is also known, being the height of a pentagon.

From Chapter 8 *Pentagon Construction Part II* we know this distance to be $\frac{\Phi\sqrt{\Phi^2+1}}{2}$ *ids.*

In Figure 15.14, the line $\overline{OM}$ is drawn perpendicular to the triangular face at its midpoint. Therefore the angle $\angle OMA$ and the angle $\angle OMX$ are right.

Note that from the above data, we can conclude that the triangles $\triangle OMA$, $\triangle OMX$, $\triangle XZO$, and $\triangle XZL$ are right.

$\overline{OM}$ can be determined because we know $\overline{OA}$ and $\overline{AM}$.

Therefore the angle $\angle MOA$ can be calculated.

With $\overline{OM}$ and $\overline{MX}$ known, $\overline{OX}$ can be determined, because triangle $\triangle OMX$ is right.

Therefore the angle $\angle MOX$ can be calculated.

Now $\angle AOX$ is known, and we can get $\angle XOZ$, because we know $\angle LOA$ is right.

Because triangle $\triangle XZO$ is right, and we know $\angle XOZ$, the angle $\angle ZXO$ can be calculated.

Now we know the angle $\angle ZXA$, it being the sum of $\angle ZXO$ and $\angle OXA$.

Now we are left with triangle $\triangle LZX$, which is right by construction.

$\overline{XZ}$ can be determined from the data in triangle $\triangle XZO$, and then $\angle ZXL$ can be calculated from triangle $\triangle LZX$.

The dihedral angle $\angle LXA$ is then the sum of $\angle ZXL$ and $\angle ZXA$.

That was quite a lot of work, but now the fun stuff is done, which for me is figuring out the geometry. Now it's just a bunch of calculations!

Our work is made easier because we can get the distances $\overline{OM}$ and $\overline{OX}$ from our previous calculations. $\overline{OM}$ is just the distance from the centroid of the i.d. to a mid-triangular face, and $\overline{OX}$ is just the distance from the centroid to any mid-edge.

We determined that $\overline{OM} = \frac{\Phi^2}{\sqrt{3}}$ *ids,* and that $\overline{OX} = \frac{\Phi\sqrt{\Phi^2+1}}{2}$ *ids.*

Let's get going!

$$\sin(MOA) = \frac{\overline{AM}}{\overline{OA}} = \frac{\frac{1}{\sqrt{3}}}{\Phi} = \frac{1}{\sqrt{3}\Phi}.$$

$$\angle MOA = 20.90515744^\circ.$$

$$\tan(\angle MOX) = \frac{\overline{MX}}{\overline{OM}} = \frac{\frac{1}{2\sqrt{3}}}{\frac{\Phi^2}{\sqrt{3}}} = \frac{1}{2\Phi^2}.$$

$$\angle MOX = 10.81231696^\circ.$$

$$\angle XOA \text{ is therefore } \angle MOA + \angle MOX = 31.71747441^\circ. \tag{15.11}$$

Well, well well. Here we recognize an angle which is a part of our good friend the Phi Right Triangle with sides divided in Mean and Extreme Ratio.

Triangle $\triangle XOA$ is not right, but triangle $\triangle XZO$ is.

Since $\angle LOA$ is right,

$$\angle LOX = \angle ZOX = \angle LOA - \angle XOA = 90° - 31.71747441° = 58.28252559°.$$

$$\text{Therefore} \angle ZXO = 31.71747441° \tag{15.12}$$

and triangle $\triangle XZO$ has sides divided in Mean and Extreme Ratio.

Now we know that $\frac{\overline{XZ}}{\overline{OZ}} = \Phi$.

$$\tan(\angle OXM) = \frac{\overline{OM}}{\overline{MX}} = \frac{\frac{\Phi^2}{\sqrt{3}}}{\frac{1}{2\sqrt{3}}} = 2\Phi^2.$$

$$\angle OXM = 79.18768304°. \tag{15.13}$$

(Note: We could also have calculated $\angle OXM$ using the following reasoning: Because triangle $\triangle OMX$ is right,

$$\angle OXM = \angle OXA = 180° - 90° - 10.81231696° = 79.18768304°).$$

$$\text{So} \angle ZXA = \angle ZXO + \angle OXM = 110.9051574°.$$

$$\text{Or more precisely, } \angle ZXA = \arctan\left(\frac{1}{\Phi}\right) + \arctan\left(2\Phi^2\right). \tag{15.14}$$

We are almost there.

If we can find $\angle ZXL$, we can finally determine $\angle LXA$.

We know $\overline{XL}$. We need to get either $\overline{LZ}$ or $\overline{XZ}$.

Lets get $\overline{LZ}$.

$\overline{LZ} = \overline{LO} - \overline{ZO}$. What is $\overline{ZO}$?

To determine that, we need $\overline{OX}$, which we already know.

Now we can write:

$$\cos(\angle ZOX) = \frac{\overline{ZO}}{\overline{OX}}, \ \overline{ZO} = \overline{OX}\left(\cos(\angle ZOX)\right).$$

We know that triangle $\triangle XZO$ is a Phi triangle and therefore

$$\cos(\angle ZOX) = \frac{1}{\sqrt{\Phi^2 + 1}}.$$

$$\overline{ZO} = \frac{\Phi\sqrt{\Phi^2 + 1}}{2}\left(\frac{1}{\sqrt{\Phi^2 + 1}}\right) = \frac{\Phi}{2} = 0.809016995. \tag{15.15}$$

Now, since $\overline{LO} = \text{radius} = \Phi \ (ids)$, then

$$\overline{LZ} = \overline{LO} - \overline{ZO} = \Phi - \frac{\Phi}{2} = \frac{\Phi}{2} \ ids.$$

Therefore $\overline{LZ} = \overline{ZO}$.

$$\sin(\angle ZXL) = \frac{\overline{LZ}}{\overline{XL}} = \frac{\frac{\Phi}{2}}{\frac{\Phi\sqrt{\Phi^2+1}}{2}} = \frac{1}{\sqrt{\Phi^2+1}}.$$

$$\angle ZXL = 31.71747441°.$$

Dihedral angle $\angle LXA = \angle ZXL + \angle ZXA$

$$= 110.9051574° + 31.71747441° = 142.6226318°. \qquad (15.16)$$

Note that

- $\overline{OX} = \overline{XL}$ and so triangle $\triangle XZO$ congruent to triangle $\triangle XZL$.
- Triangle $\triangle XLO$ is isosceles.
- $\frac{\overline{XL}}{\overline{LZ}}$ also equals Φ.

15.8. Centroid Distances

Let's collect some information we have already calculated:

$$\text{Distance from centroid to a vertex} = \text{radius} = \Phi \, ids.$$

$$\text{Distance from centroid to mid-pentagonal face} = \frac{\Phi^2}{\sqrt{\Phi^2+1}} \, ids.$$

$$\text{Distance from centroid to mid-triangular face} = \frac{\Phi^2}{\sqrt{3}} \, ids.$$

$$\text{Distance from centroid to mid-edge} = \frac{\Phi\sqrt{\Phi^2+1}}{2} \, ids.$$

Or, in decimals,

$$\text{Distance from centroid to a vertex} = \text{radius} = 1.618033989 \, ids.$$

$$\text{Distance from centroid to mid-pentagonal face} = 1.376381921 \, ids.$$

$$\text{Distance from centroid to mid-triangular face} = 1.511522629 \, ids.$$

$$\text{Distance from centroid to mid-edge} = 1.538841769 \, ids.$$

15.9. Planar Distances

Now let's calculate the separation of the various planes in the icosadodecahedron.

We want to get the distances between E_0, F_0, G_0, I_0, and H_0.

Let's first get $\overline{E_0 F_0}$. Refer to Figure 15.15.

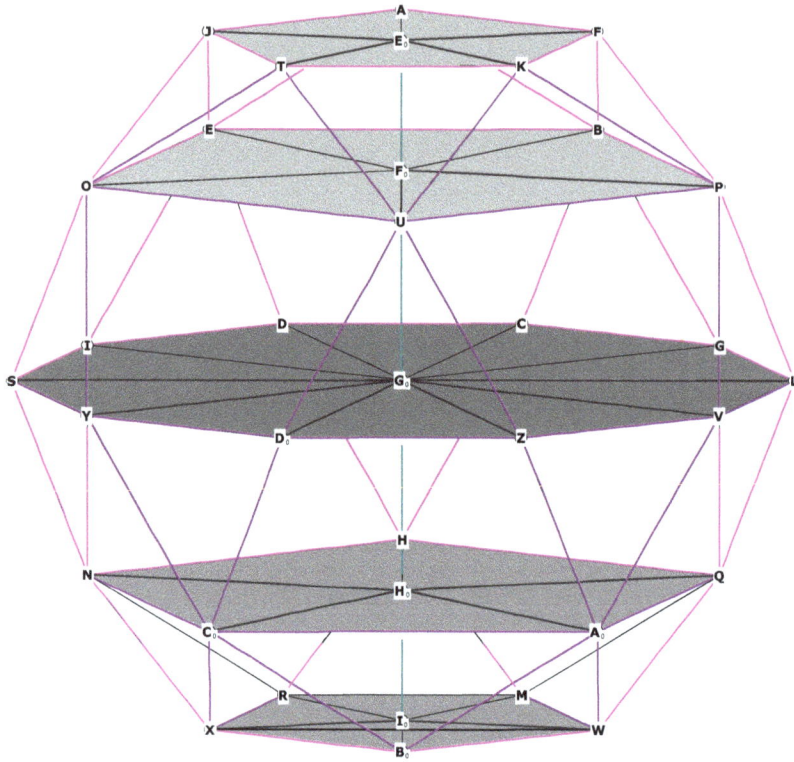

Figure 15.15: Four of the internal pentagons and 1 internal decagon

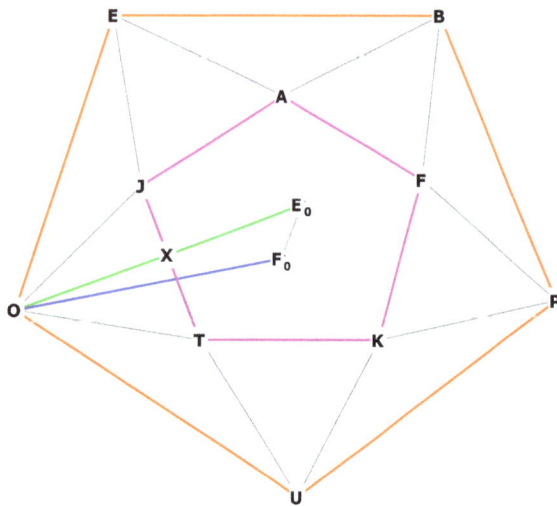

Figure 15.16: Top view, refer to Figure 15.15

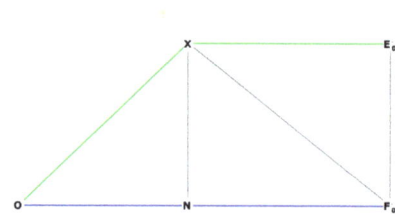

Figure 15.17: The distance between the first 2 pentagonal planes ($\overline{E_0 F_0}$) of the icosadodecahedron

Note that although $\overline{OXE_0}$ appears to be a straight line in Figure 15.16, it is actually the dihedral angle of the solid.

The top pentagonal plane on the i.d. is the plane AJTFK, in purple in Figure 15.16 and the top plane in Figures 15.15, 15.16, and 15.17 on p. 142. In Figure 15.16, the large pentagonal plane UPBEO is marked in orange.

The point E_0 is the center of the small pentagon, F_0 is the center of the large pentagon.

AJTFK is on top of UPBEO.

The angle $\angle F_0 E_0 X$ is right. EoFo is the line perpendicular to the plane AJTFK from the center of the plane UPBEO, and is part of the diameter of the sphere which runs through the centroid G_0.

The angle $\angle OF_0 E_0$ is also right, the line $\overline{E_0 F_0}$ is perpendicular to the plane UPBEO.

$\overline{OX}$ and $\overline{XE_0}$ can be easily calculated.

$\overline{OX}$ is just the height of an equilateral triangle, $\frac{\sqrt{3}}{2} ids$, and $\overline{XE_0}$ is the distance, in a pentagon, from the center to any mid-edge. We know from Chapter 8 *Construction of the Pentagon Part II* that this distance is $\frac{\Phi^2}{2\sqrt{\Phi^2+1}} ids$.

The distance $\overline{OF_0}$ is just the distance from the center of the large pentagon to one of its vertexes. From *Construction of the Pentagon Part II* we know this distance to be $\frac{\Phi}{\sqrt{\Phi^2+1}} \cdot$(side of the large pentagon).

We may draw a perpendicular bisector from X to N on the line $\overline{OF_0}$.

Triangles $\triangle E_0 XN$ and $\triangle XNF_0$ are both right.

We now have a rectangle XNF_0E_0 with four right angles at the corners, and $\overline{XN} = \overline{E_0 F_0}$, $\overline{XE_0} = \overline{NF_0}$.

Now we can find the distance $\overline{ON}$, for it is just $\overline{OF_0} - \overline{NF_0}$.

Then we can work with triangle $\triangle ONX$ to find $\overline{XN}$ and we have $\overline{E_0 F_0}$.

First we must find $\overline{OF_0}$ in terms of the side of the icosadodecahedron.

Remember that $\overline{OF_0}$ is $\frac{\Phi}{\sqrt{\Phi^2+1}}$ (side of the large pentagon). But the side of the large pentagon is just the diagonal of any the pentagonal faces of the i.d.

Therefore the side of the large pentagon is $\Phi\,(ids)$.

We can then write

$$\overline{OF_0} = \frac{\Phi}{\sqrt{\Phi^2 + 1}} \left(\Phi\,ids\right) = \frac{\Phi^2}{\sqrt{\Phi^2 + 1}}\, ids.$$

(Note that this distance is precisely the distance from the centroid to the mid-face of any of the pentagonal faces of the icosadodecahdron!)

$$\overline{NF_0} = \overline{XE_0} = \frac{\Phi^2}{2\sqrt{\Phi^2 + 1}}\, ids.$$

$$\text{Then, } \overline{ON} = \overline{OF_0} - \overline{NF_0} = \frac{\Phi^2}{\sqrt{\Phi^2 + 1}}\, ids - \frac{\Phi^2}{2\sqrt{\Phi^2 + 1}}\, ids.$$

$$= \frac{\Phi^2}{2\sqrt{\Phi^2 + 1}}\, ids. \tag{15.17}$$

From this calculation we gather that $\overline{ON} = \overline{NF_0}$.

$$\overline{XN}^2 = \overline{OX}^2 - \overline{ON}^2$$

$$= \frac{3}{4} - \frac{\Phi^4}{4\left(\Phi^2 + 1\right)}$$

$$= \frac{3\left(\Phi^2 + 1\right) - \Phi^4}{4\left(\Phi^2 + 1\right)}$$

$$= \frac{4}{4\left(\Phi^2 + 1\right)} = \frac{1}{\Phi^2 + 1}.$$

$$\overline{XN} = \overline{E_0 F_0} = \frac{1}{\sqrt{\Phi^2 + 1}} \, ids.$$

Therefore the distance between the first two pentagonal planes is

$$\frac{1}{\sqrt{\Phi^2 + 1}} \, ids. \tag{15.18}$$

Now we can easily find $\overline{F_0 G_0}$, the distance between the large pentagonal plane and the plane of the decagon which contains the centroid.

This distance is just $\overline{E_0 G_0} - \overline{E_0 F_0}$. $\overline{E_0 G_0}$ is the distance from the centroid to mid-face of pentagon, which we found above to be $\frac{\Phi^2}{\sqrt{\Phi^2+1}} \, ids$. So

$$\overline{F_0 G_0} = \frac{\Phi^2}{\sqrt{\Phi^2 + 1}} ids - \frac{1}{\sqrt{\Phi^2 + 1}} ids$$

$$= \frac{\Phi^2 - 1}{\sqrt{\Phi^2 + 1}} \, ids$$

$$= \frac{\Phi}{\sqrt{\Phi^2 + 1}} \, ids. \tag{15.19}$$

15.9.1. The relationship between $\overline{E_0 F_0}$, $\overline{F_0 G_0}$ and $\overline{E_0 G_0}$

$$\frac{\overline{E_0 G_0}}{\overline{F_0 G_0}} = \frac{\frac{\Phi^2}{\sqrt{\Phi^2+1}}}{\frac{\Phi}{\sqrt{\Phi^2+1}}} = \frac{\Phi^2}{\Phi} = \Phi! \tag{15.20}$$

Therefore the radius $\overline{E_0 G_0}$ is divided in Mean and Extreme Ratio at F_0.

15.9.2. TABLE OF RELATIONSHIPS OF THE DISTANCES BETWEEN THE PENTAGONAL PLANES

Table 15.1: Relationships of the Distances between the Internal Pentagonal-Decagon Planes of the i.d. (Let $E_0 F_0 = 1$). (Graph reads vertically, by column)

E_0	E_0
1	
F_0	Φ^2
Φ	
G_0	G_0
Φ	
H_0	Φ^2
1	
I_0	I_0

15.9.3. DISTANCES BETWEEN THE TRIANGULAR PLANES AND THE HEXAGONAL PLANE WHICH RUNS THROUGH THE CENTROID

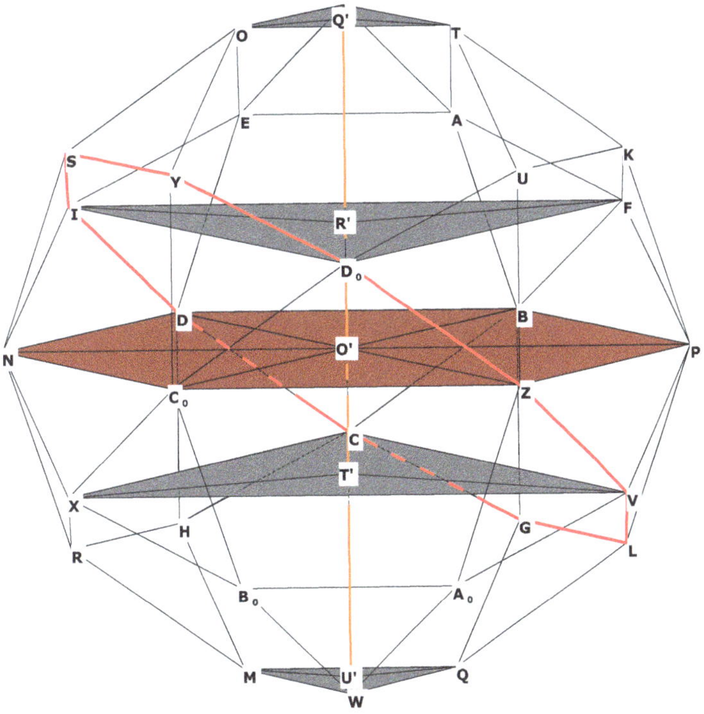

Figure 15.18: hexagonal and triangular internal planes

The distance $\overline{R'Q'}$ and $\overline{T'U'}$.

The large equilateral triangle $\triangle D_0IF$ in Figure 15.18 has sides equal to the distance $\overline{D_0I}$ in Figure 15.19. Figure 15.19 shows the length of one of the sides of the large equilateral triangular plane D_0IF with respect to one of the large "great circle" decagons marked in orange. The other two legs of this large equilateral triangle are each part of different decagons. Each of the four triangular planes in Figure 15.18 is equilateral, which can easily be seen by constructing a 3-D model.

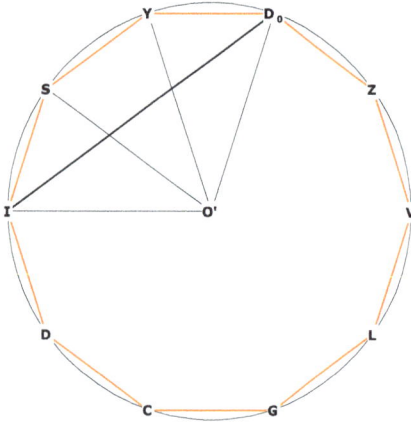

Figure 15.19: Showing the length of the sides of the large triangular planes with respect to one of the great circle decagon sides

From *Decagon* in Appendix D we know the edge length of each side of triangle $\triangle D_0IL$ is Φ^2 ids. This is seen in Appendix D as the distance $\overline{DA}$, and in Figure 15.19 as $\overline{D_0I}$.

In Figure 15.18, R' is the center of the triangular face D_0IF. T' is the center of the triangular face CXV, and O' is the centroid. We want to find $\overline{R'T'}$, the distance between the two triangular planes. We will first find $\overline{O'R'}$.

The solution is quite simple, actually. The distance $\overline{D_0R'}$ is known. $\overline{D_0R'}$ is the distance from the center of an equilateral triangle to an i.d. vertex. $\overline{O'R}$ is also known. $\overline{O'R}$ is just the distance from the centroid to a vertex of the i.d. R' lies directly above O'. Therefore the triangle $\triangle O'R'D'$ is right and we may write

$$\overline{O'R'}^2 = \overline{O'D_0}^2 - \overline{R'D_0}^2.$$

$\overline{O'D_0} = \Phi$ *ids*. From *Equilateral Triangle* we know that $\overline{R'D_0} = \frac{1}{\sqrt{3}}$ (side of triangle), in this case, Φ^2 *ids*.

$$\overline{O'R'}^2 = \Phi^2 \, ids^2 - \left(\frac{\Phi^2}{\sqrt{3}}\right)^2 ids^2$$

$$= \frac{3\Phi^2 - \Phi^4}{3} \, ids^2$$

$$= \frac{3\Phi^2 - 3\Phi^2 + 1}{3}$$

$$= \frac{1}{3} \, ids^2.$$

$$\overline{O'R'} = \frac{1}{\sqrt{3}} \, ids = 0.577350269 \, ids. \tag{15.21}$$

Because the analysis is precisely the same for $\overline{O'T'}$, the distance $\overline{O'T'}$ also equals $\frac{1}{\sqrt{3}}\,ids$, and

$$\overline{R'T'} = \frac{2}{\sqrt{3}}\,ids = 1.154700538\,ids. \tag{15.22}$$

The Distance $\overline{Q'R'}$

This is the distance between the triangular face OJT and the plane D_0IF in Figure 15.18. The analysis is the same as above, except that the side of the triangular plane OJT is now ids, instead of $\Phi^2\,(ids)$. (Why? Because OJT is a face of the i.d.) Therefore $\overline{Q'R'}$ is the distance from the centroid O' to the mid-triangular face.

$$\overline{Q'R'} = \overline{O'Q'} - \overline{O'R'}$$

$$\overline{Q'R'} = \frac{\Phi^2}{\sqrt{3}}\,ids - \frac{1}{\sqrt{3}}\,ids$$

$$= 0.934172359\,ids. \tag{15.23}$$

$$\frac{\overline{Q'R'}}{\overline{O'R'}} = \frac{\frac{\Phi}{\sqrt{3}}}{\frac{1}{\sqrt{3}}} = \Phi. \tag{15.24}$$

Therefore $\overline{O'Q'}$ is divided in Mean and Extreme Ratio at R', and the triangular planes and the hexagonal plane ($\sqrt{3}$ geometry) are divided in Phi ratio ($\sqrt{5}$ geometry).

15.9.4. TABLE OF RELATIONSHIPS OF THE DISTANCES BETWEEN THE TRIANGULAR AND HEXAGONAL PLANES

Table 15.2: Relationships of the Distances between the Internal Pentagonal-Decagon Planes of the i.d. (Let $O'R' = 1$). (Graph reads vertically, by column)

Q'	Q'
Φ	
R'	Φ^2
1	
O'	O'
1	
T'	Φ^2
Φ	
U'	U'

15.10. The Angle Any Hexagonal Plane Makes With the Plane of the Decagon

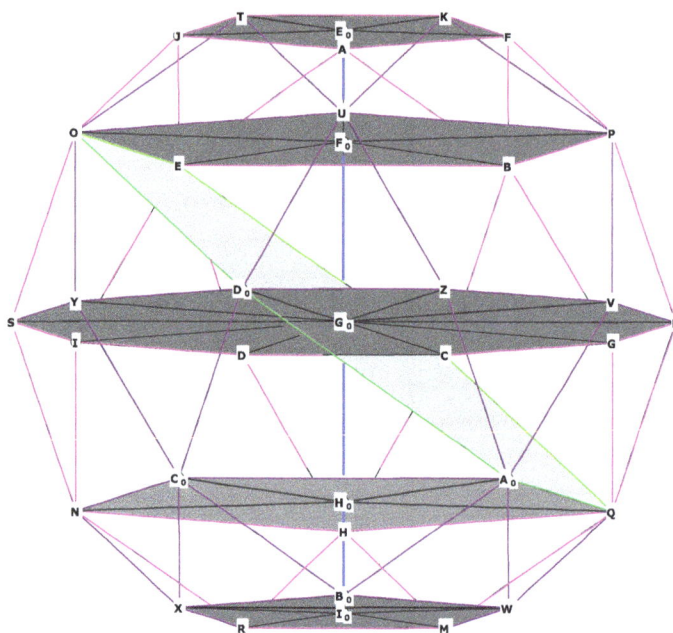

Figure 15.20: Figure 15.6, repeated

In Figure 15.20, the hexagonal plane is highlighted, and can be seen by observing the plane of the hexagon as it intersects the plane of the decagon at the centroid, G_0.

Notice that $\angle OG_0S$ is the angle of the hexagonal plane as it intersects with the central decagon, and that is it also a central angle of the i.d.

$$\text{This angle is } \frac{360°}{10} = 36°. \tag{15.25}$$

15.11. Conclusions

What have we learned about the icosadodecahedron?

- The icosa-dodecahedron is pentagonally based.
- When the icosahedron is seated on one of its triangular faces, there are 4 internal equilateral triangular planes and one internal decagon plane, all separated in Phi ratio.
- When the icosadodecahedron is seated on one of its pentagonal planes, there are 4 internal pentagonal planes and one decagon plane, all separated in Phi ratio.
- All of the radii from the centroid G_0 form Golden Triangles with any two adjacent vertices.

15.12. Icosa-dodecahedron Reference Charts

Table 15.3: Volume and Surface Area

Volume in terms of s	Volume in Unit Sphere	Surface Area in terms of s	Surface Area in Unit Sphere
13.83552595 s³	3.266124627 r³	29.30598285 s²	11.19388937 r²

Table 15.4: Angles

Central Angle:	Dihedral Angle:	Surface Angles:	
36°	142.6226318°	triangular face	pentagonal face
		60°	108°

Table 15.5: Centroid Distances

Centroid To Vertex:	Centroid To Mid-edge:	Centroid To Mid-triangular face:	Centroid To Mid-pentagonal face:
1.0 r	0.951056516 r	0.934172359 r	0.850650809 r
1.6180339898 (Φ) s	1.538841769 s	1.511522629 s	1.376381921 s

Table 15.6: Side to Radius

Side / radius
0.618033989 $\left(\frac{1}{\Phi}\right)$

Chapter 16

The Rhombic Triacontahedron

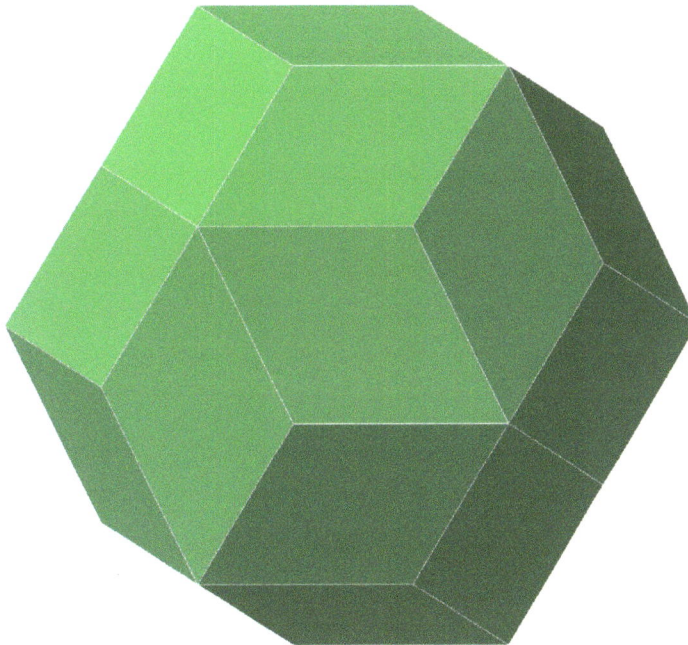

Figure 16.1: The Rhombic Triacontahedron

The Rhombic Triacontahedron is an extremely fascinating polyhedron. It is built around the dodecahedron, and like the dodecahedron, it has many phi relationships within it. Remarkably, this polyhedron contains all five of the Platonic Solids directly on its vertices, and shows the proper relationship between them.

The Rhombic Triacontahedron (hereinafter referred to as r.t.) has 30 faces, 60 edges, and 32 vertices.

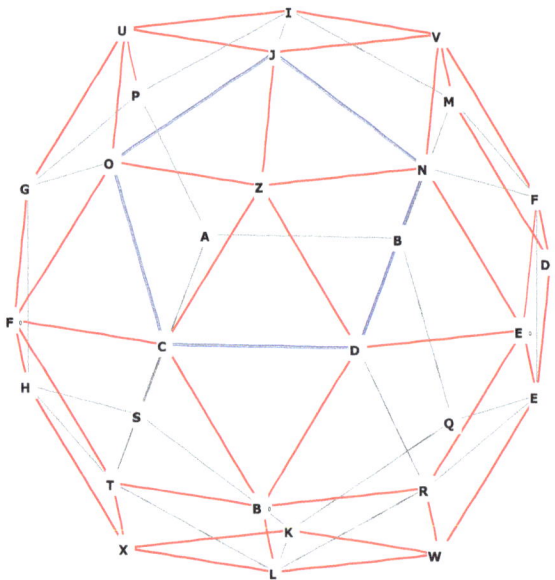

Figure 16.2: The front portion of the rhombic triacontahedron (red) with the sides of the dodecahedron (gray)

Notice the dodecahedral pentagon CDNJO, then notice Z. Z is raised off the pentagon (a distance we shall find later on) and Z is connected to all five vertices of the pentagonal face of the dodecahedron. This can be seen even more clearly at pentagon GPIJO, where U rises off its face. The pentagon GPIJO has all of its vertices connected to U. Of course, we could have drawn this figure without the dodecahedron faces, which is obviously not part of the r.t., but it helps for clarity.

Notice that the faces of the r.t. are diamond-shaped, somewhat like the rhombic dodecahedron, but these faces are longer and skinnier. Look at the r.t. face UOZJ at the upper left. U and Z are the long-axis vertices. Notice that the line between O and J, the short-axis vertices, form one of the edges (sides) of the dodecahedron.

The r.t. is more clearly spherical than any of the polyhedra we have studied so far.

In Figure 16.3 below, we see that the rhombic triacontahedron has internal pentagonal planes just like the icosahedron and the dodecahedron!

Note the large highlighted pentagons. The lengths of all of the large internal pentagon sides are precisely the long axis of every r.t. face! For example, look at $\overline{VZ}$ and $\overline{ZF_0}$ in the top pentagonal plane and $\overline{E_0B_0}$ in the lower pentagonal plane.

The r.t. has a circumsphere and an inner sphere. The circumsphere goes around all 12 of the vertices that rise off the 12 faces of the dodecahedron.

The diameter of the circumsphere is $\overline{UW}$. $\overline{UW}$ is in green in Figure 16.3. O' is the centroid.

The inner sphere touches all of the 20 vertices of the dodecahedron. The diameter of the inner sphere would be, for example, $\overline{IL}$. $\overline{UW} > \overline{IL}$.

This leads us to think that possibly, the relationship between U, G_0, O', H_0, and W will be similar to those of the dodecahedron and the icosahedron. Later on, we will see that this is indeed the case.

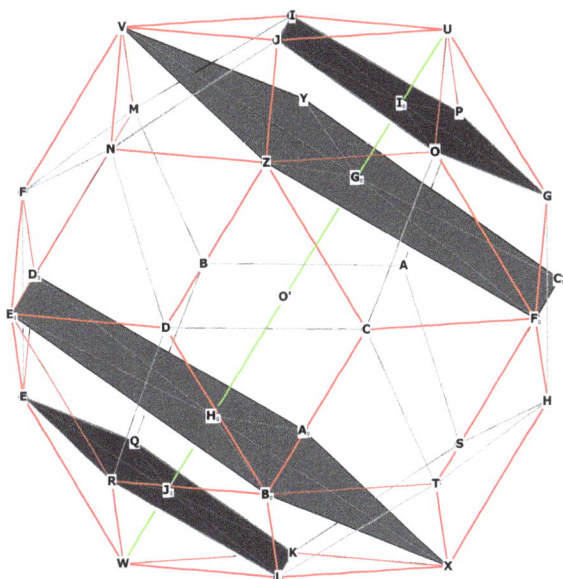

Figure 16.3: The internal pentagonal planes of the r.t. This figure is rotated 180° from Figure 16.2

Notice also that the rhombic triacontahedron contains on its vertices not only a dodecahedron, but an icosahedron as well! (See Figure 16.4.) Notice that the 12 vertices that rise off the dodecahedron faces provide the 12 vertices of the icosahedron. Remember that the dual of the dodecahedron is the icosahedron and that the dual is formed by taking points at the center of the faces.

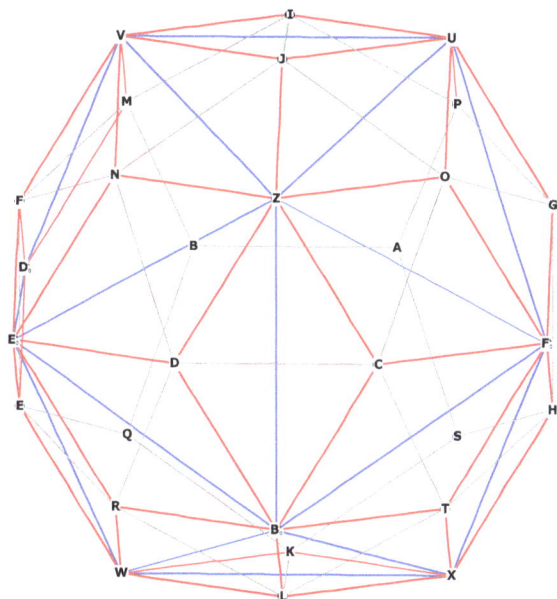

Figure 16.4: the 12 vertices of the icosahedron (blue) within the rhombic triacontahedron. The dodceahedron is shown in dark green.

The sides of the icosahedron are the long axes of the r.t. faces.

The sides of the dodecahedron are the short axes of the r.t. faces.

Therefore the rhombic triacontahedron is just the combination of the dodecahedron with its dual, the icosahedron.

The rhombic triacontahedron is the model nature uses to demonstrate the true relationship between the side of the icosahedron and the side of the dodecahedron. We will see later on that these relationships are based on Φ.

Note also that when a cube is inscribed within 8 of the 12 vertices belonging to the dodecahedron within the rhombic triacontahedron, the lengths of the cube sides are equal to the long axis of any of the rhombic triacontahedron rhombi. Furthermore, the sides of the cube are equal to any of the diagonals of the pentagonal faces of the dodecahedron (see Figure 16.5).

Because the tetrahedron (one tetrahedron in green, the other in purple) and the octahedron (in orange) can be inscribed within the cube, the rhombic triacontahedron shows the precise relationship between the Platonic Solids! (See Figure 16.6.)

The Rhombic Triacontahedron therefore elegantly describes the nesting of the five Platonic Solids: icosahedron, dodecahedron, cube, tetrahedron, octahedron. When the sides of the octahedron are divided in Mean and Extreme (Phi) Ratio, another icosahedron is formed (see figure 16.7). This begins the process all over again, and shows that the five nested Platonic Solids may not only grow and contract to infinity, but do so in a perfectly harmonious way.

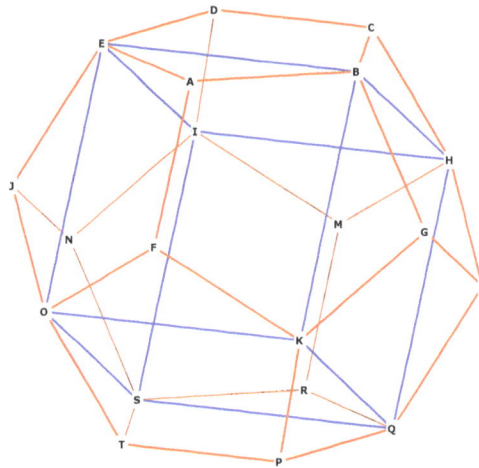

Figure 16.5: The sides of the cube, in blue, are also the diagonals of the pentagonal faces of the dodecahedron within the rhombic triacontahedron.

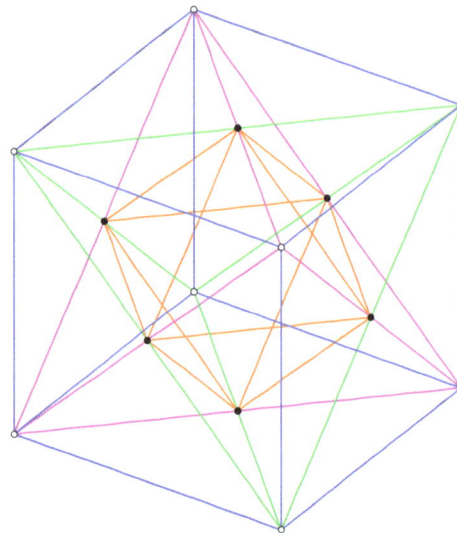

Figure 16.6: The cube, two interlocking tetrahedrons, and the octahedron inside a cube

Interestingly, the icosahedron is formed within the octahedron by dividing each edge of the octahedron in Phi Ratio. Here again is an elegant link between root 2 and root 3 geometry, and Phi.

16.1. Volume of the Rhombic Triacontahedron

As before we use the pyramid method. There are 30 faces, so there are 30 pyramids. Imagine a point at the very center of the r.t. If you connect that point up with one of the faces, you will have a pyramid that looks like Figure 16.8.

In Figure 16.8, $\overline{O'Z}$ is the height of the pyramid. $\overline{O'U} = \overline{O'V}$ = radius of outer sphere which touches the 12 r.t. vertices that are raised off the center of the dodecahedron face. $\overline{O'I} = \overline{O'J} =$

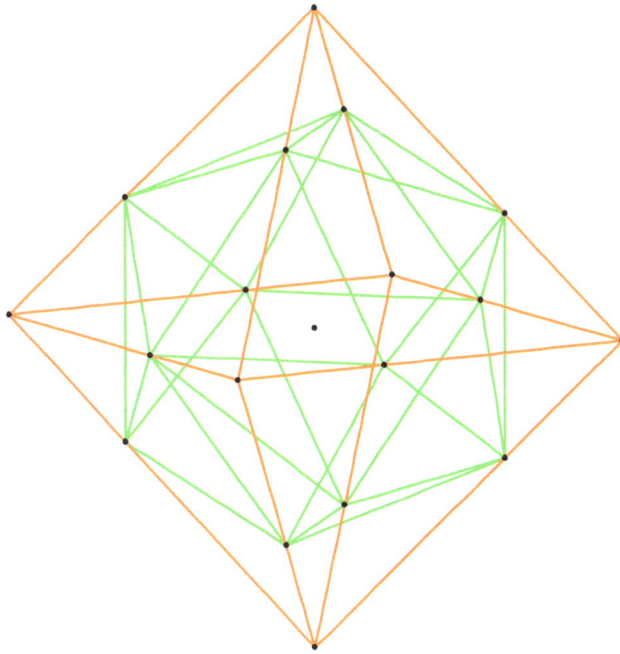

Figure 16.7: Icosahedron in octahedron

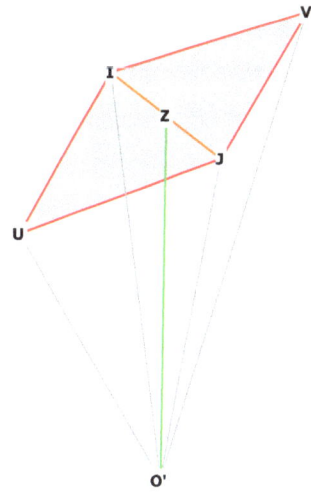

Figure 16.8: One of the 30 pyramids of the rhombic triacontahedron

radius of inner sphere which touches all 20 of the short-axis r.t. vertices, which are also the vertices of the dodecahedron.

We want to find the volume of the r.t. in terms of the r.t. side. However, we don't know the length of the r.t side! But we do know the length of the side of the dodecahedron, in terms of a unit sphere which touches all the vertices of the dodecahedron. So let's try to get the r.t. side (hereinafter referred to as rts) in terms of the dodecahedron side.

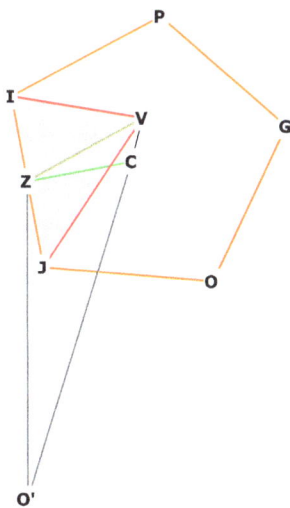

Figure 16.9

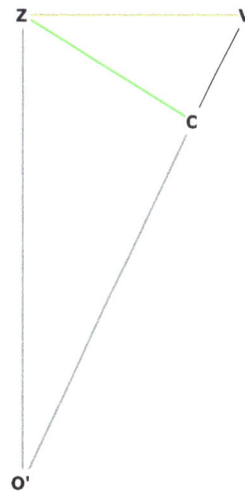

Figure 16.10

Figures 16.9 and 16.10 show the following important data:

- V is one of the 12 raised vertices of the r.t. off the face of the dodecahedron.
- C is the center of the dodecahedron face. C lies in the plane of IJOGP, and is directly below V.
- Z is the center of the r.t. face IUJV (see Figure 4), and also the mid-edge of the side of the dodec.
- IVJ (in red) is one-half of the r.t. diamond face IUJV (see Figure 4)
- $\overline{O'Z}$ is the distance from centroid to mid-edge of the dodec face. This is the height of the r.t. pyramid.
- $\overline{ZV}$ is the distance from mid-edge of dodecahedron to the raised vertex V off the dodecahedron face.
- $\overline{ZC}$ is the distance from mid-edge of dodecahedron to center of dodecahedron face.
- $\overline{CV}$ is the distance of V off the center of the dodecahedron face.

Figure 16.10 shows that $\overline{O'CV}$ is a straight line, so that $\angle ZO'C = \angle ZO'V$, $\triangle O'ZV$ is right, and $\triangle O'CZ$ is right.

From this data we can show that the triangles $\triangle O'ZV$, $\triangle O'CZ$, and $\triangle CZV$ are similar by angle-side-angle.

First we show the triangle $\triangle O'ZV$ and $\triangle O'CZ$ are similar:

- $\triangle O'ZV$ is right, so is $\triangle O'CZ$.
- $\overline{O'Z}$ is common to both triangles.
- $\angle ZO'V = \angle ZO'C$.

Therefore both triangles are similar by ASA.

Now we show that triangle $\triangle CZV$ is similar to triangle $\triangle O'ZV$ by angle-side-angle.

- $\angle ZCV$ and $\angle O'ZV$ are right.
- $\overline{ZV}$ is common to both triangles.
- $\angle O'VZ = \angle CVZ$.

Therefore both triangles are similar by angle-side-angle.

With this information we can determine $\overline{CV}$, the distance of the vertex V off of the dodec face, and $\overline{ZV}$, which will enable us to get the side of the r.t. in terms of the side of the dodecahedron.

Because all 3 triangles are similar, we can write the following relationship:

$$\frac{\overline{CZ}}{\overline{O'C}} = \frac{\overline{CV}}{\overline{CZ}}.$$

The distances $\overline{O'Z}$, $\overline{O'C}$, and $\overline{CZ}$ are known. From Chapter 11 *Dodecahedron* it is known that

$$\overline{CZ} = \frac{\Phi^2}{2\sqrt{\Phi^2+1}}\, ds. \text{(ds means the dodecahedron side)}$$

$$\overline{O'C} = \frac{\Phi^3}{2\sqrt{\Phi^2+1}}\, ds.$$

$$\overline{O'Z} = \frac{\Phi^2}{2}\, ds = h, \text{the height of the r.t. pyramid.}$$

Therefore,

$$\frac{\overline{CV}}{\overline{CZ}} = \frac{\frac{\Phi^2 s}{2\sqrt{\Phi^2+1}}}{\frac{\Phi^3 s}{2\sqrt{\Phi^2+1}}} = \frac{1}{\Phi}!$$

$$\overline{CV} = \frac{1}{\Phi}\left(\overline{CZ}\right) = \frac{\Phi}{2\sqrt{\Phi^2+1}}\,ds = 0.425325404\,ds.$$

Now we have $\overline{CV}$, the distance from the plane of the dodecahedron to the rhombic triacontahedron "cap" over the dodecahedron face, in terms of the dodecahedron side.

16.1.1. EDGE LENGTH OF THE RHOMBIC TRIACONTAHEDRON

Now we need to find $\overline{ZV}$, so that we can get $\overline{IV}$, the length of the side or edge of the rhombic triacontahedron. We can write the following relationship:

$$\frac{\overline{CZ}}{\overline{ZV}} = \frac{O'C}{O'Z} = \frac{\frac{\Phi^3}{2\sqrt{\Phi^2+1}}}{\frac{\Phi^2}{2}} = \frac{\Phi}{\sqrt{\Phi^2+1}}\,ds.$$

$$\overline{ZV} = \frac{\sqrt{\Phi^2+1}}{\Phi}\left(\overline{CZ}\right)$$

$$= \frac{\sqrt{\Phi^2+1}}{\Phi}\left(\frac{\Phi^2}{2\sqrt{\Phi^2+1}}\right)$$

$$= \frac{\Phi}{2}\,ds = 0.809016995\,ds. \tag{16.1}$$

Refer back to Figures 16.8 and 16.9. Now that we have $\overline{ZV}$, we can find $\overline{IV}$, the side of the r.t.

The triangle $\triangle IVJ$ in Figures 16.8 and 16.9 is one–half of an r.t face. We have already calculated $\overline{ZV}$, and we know that $\overline{IZ}$ is just one–half the side of the side of the dodecahedron, ds. We also know that the angle $\angle IZV$ is right by construction.

Therefore, by the Pythagorean Theorem,

$$\overline{IV}^2 = \overline{IZ}^2 + \overline{ZV}^2 = \frac{1}{4} + \frac{\Phi^2}{4} = \frac{\Phi^2+1}{4}\,ds^2.$$

$$\overline{IV} = rts = \frac{\sqrt{\Phi^2+1}}{2}\,ds,\text{ and}$$

$$ds = \frac{2}{\sqrt{\Phi^2+1}}\,rts = 1.0514562224\,rts. \tag{16.2}$$

Here we have established an important fact: we have related the side of the rhombic triacontahedron, or rts, to the side of the dodecahedron and, therefore, to the radius of the unit sphere which encloses all 20 vertices of the dodecahedron and, in turn, the short axis vertices of the rhombic triacontahedron.

We can now describe the distance of any vertex of the r.t. off the plane of the dodecahedron, $\overline{CV}$, in terms of the side of the r.t. We may now write

$$\overline{CV} = \frac{\Phi}{2\sqrt{\Phi^2+1}}\,ds = \frac{\Phi}{2\sqrt{\Phi^2+1}}\left(\frac{2}{\sqrt{\Phi^2+1}}\,rts\right) = \frac{\Phi}{\Phi^2+1}\,rts.$$

$$\overline{CV} = 0.447213596\,rts. \tag{16.3}$$

16.1.2. RADIUS OF OUTER CIRCUMSPHERE

Refer to Figure 16.11 below. Let's now find the radius of the outer sphere of the r.t. in terms of the side of the r.t. itself. Remember that the outer sphere touches all 12 long-axis vertices of the r.t., and that these 12 vertices are the vertices of an icosahedron. Refer to figure 16.11

$r_{\text{outer}} = \overline{O'V} = \overline{O'C} + \overline{CV}$. We need to convert these distances so that they are related to rts, not ds.

$$\overline{O'C} = \frac{\Phi^3}{2\sqrt{\Phi^2+1}}\, ds$$
$$= \frac{\Phi^3}{2\sqrt{\Phi^2+1}} \left(\frac{2}{\sqrt{\Phi^2+1}}\, rts \right)$$
$$= \frac{\Phi^3}{\Phi^2+1}\, rts.$$
$$\overline{CV} = \frac{\Phi}{2\sqrt{\Phi^2+1}}\, ds = \frac{\Phi}{2\sqrt{\Phi^2+1}} \left(\frac{2}{\sqrt{\Phi^2+1}}\, rts \right)$$
$$= \frac{\Phi}{\Phi^2+1}\, rts. \tag{16.4}$$

So

$$r_{outer} = \overline{O'V} = \frac{\Phi^3}{\Phi^2+1}\, rts + \frac{\Phi}{\Phi^2+1}\, rts = \frac{\Phi\left(\Phi^2+1\right)}{\Phi^2+1}\, rts = \Phi.$$
$$\text{Therefore, } r_{outer} = \overline{O'V} = \Phi\left(rts\right). \tag{16.5}$$
$$rts = \frac{r_{outer}}{\Phi}. \tag{16.6}$$

In the Rhombic Triacontahedron, the relationship between the side and the radius of the enclosing sphere is Phi. Interestingly, this is precisely what we found with the icosa–dodecahedron, the dual of the rhombic triacontahedron. Great minds think alike, as they say.

(BTW, what did the rhombic triacontahedron say to the icosadodecahedron? "You're Phi–ntastic." OK, really bad joke, but I couldn't resist!)

16.1.3. INNER SPHERE RADIUS OF RHOMBIC TRIACONTAHEDRON

Refer to Figure 16.11. What is r_{inner}? This is just the unit sphere which touches all 20 vertices of the dodecahedron. We know from *Dodecahedron* that $r_{inner} = \frac{\sqrt{3}\Phi}{2}\, ds$.

Converting this to the side of the r.t. we have:

$$r_{inner} = \frac{\sqrt{3}\Phi}{2}\, ds$$
$$= \frac{\sqrt{3}\Phi}{2} \left(\frac{2}{\sqrt{\Phi^2+1}}\, rts \right)$$
$$= \frac{\sqrt{3}\Phi}{\sqrt{\Phi^2+1}}\, rts = 1.47337042\, rts. \tag{16.7}$$

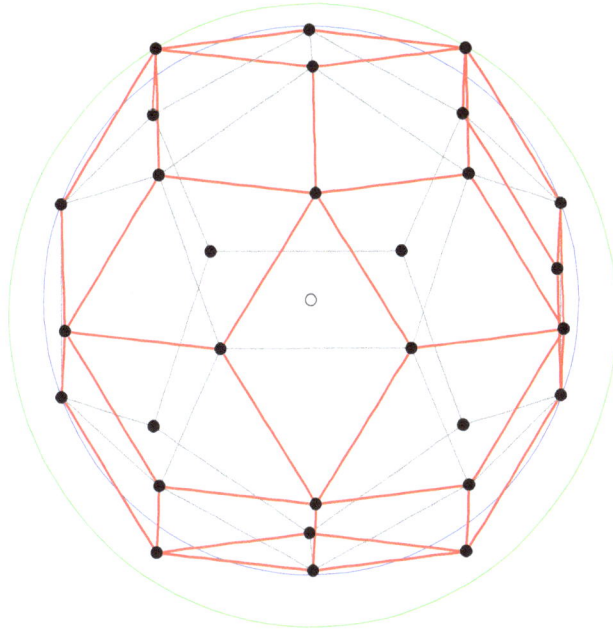

Figure 16.11: showing inner and outer circumspheres of rhombic triacontahedron. The outer sphere (r_{outer}) touches the vertices of the r.t. on the long axis. The inner sphere (r_{inner}) touches the vertices of the r.t. on the short axis, each of which vertices forms a dodecahedron

Note how the $\sqrt{3}$ appears in the numerator. The dodecahedron is the link between the $\sqrt{3}$ geometry of the cube, tetrahedron and octahedron, and Φ, which appears over and over again in more complex polyhedra.

Now we have enough information to calculate the volume of the rhombic triacontahedron in terms of it's own side. Refer back to Figures 16.8 and 16.9 for diagrams.

We can calculate the area of the r.t. face, because we know $\overline{IJ}$ and $\overline{ZV}$, which we can use to give us the area of one-half the face of the r.t. We see from Figure 16.8 that $\overline{O'Z}$, the height of the r.t. pyramid, is just the distance from the centroid to the mid-edge of any of the sides of the dodecahedron. We know from *Dodecahedron* that this distance is $\frac{\Phi^2}{2} ds$. We have to do some conversions of these values first, to get them all in terms of rts.

To get the area of the r.t. face, divide it into two identical triangles at the short axis $\overline{IJ}$, find the area of one triangle, and multiply by 2.

The area of any triangle is $\frac{1}{2}$ (base) (height), so the area of the r.t. face is twice this value.

$\overline{IJ}$ is just the side of the dodecahedron, and it is the base of our triangle.

$$\overline{IJ} = ds = \frac{2}{\sqrt{\Phi^2+1}}\,rts. \tag{16.8}$$

$$\overline{ZV} = \text{height of triangle} = \frac{\Phi}{2}\,ds = \frac{\Phi}{2}\left(\frac{2}{\sqrt{\Phi^2+1}}\,rts\right) = \frac{\Phi}{\sqrt{\Phi^2+1}}\,rts.$$

$$Area_{\text{1 triangle}} = \frac{1}{2}\left(\frac{2}{\sqrt{\Phi^2+1}}\right)\left(\frac{\Phi}{\sqrt{\Phi^2+1}}\right) = \frac{\Phi}{\Phi^2+1}\,rts^2.$$

$$Area_{\text{1 diamond r.t. face}} = \frac{2\Phi}{\Phi^2+1}\,rts^2.$$

$$Volume_{\text{1 r.t. pyramid}} = \frac{1}{3}\,(\text{area of face})\,(\text{pyramid height}).$$

$$= \frac{1}{3}\,(\text{area of face})\,(\overline{O'Z})$$

$$= \frac{1}{3}\left(\frac{2\Phi}{\Phi^2+1}\,rts^2\right)\left(\frac{\Phi^2}{2}\,ds\right)$$

$$= \frac{1}{3}\left(\frac{2\Phi}{\Phi^2+1}\,rts^2\right)\left(\frac{\Phi^2}{2}\right)\left(\frac{2}{\sqrt{\Phi^2+1}}\,rts\right)$$

$$= \frac{1}{3}\left(\frac{2\Phi}{\Phi^2+1}\,rts^2\right)\left(\frac{\Phi^2}{\sqrt{\Phi^2+1}}\,rts\right)$$

$$= \frac{2\Phi^3}{3\left(\Phi^2+1\right)^{3/2}}\,rts^3. \tag{16.9}$$

There are 30 pyramids for 30 faces so total volume V_{total} is:

$$Volume_{\text{total}} = 30\left(\frac{2\Phi^3}{3\left(\Phi^2+1\right)^{3/2}}\,rts^3\right)$$

$$= \frac{20\Phi^3}{\left(\Phi^2+1\right)^{3/2}}\,rts^3 = 12.31073415\,rts^3. \tag{16.10}$$

The volume of the r.t. can be calculated another way. I include this calculation not because it's necessary, but to show that there is more than one way to get the answer. Math and geometry is more about the ability to think with the material than a series of boring "plug-in-the-numbers" exercises.

Since the r.t. is built upon the dodecahedron, the r.t. volume is just the volume of the dodecahedron + the extra volume of all of the little 12 pentagonal pyramids formed from the raised vertices off the 12 pentagonal faces of the dodecahedron. To see this, check Figure 16.2 again and look at U-IJOGP or V- IJNFM.

The volume of the dodecahedron is, from Chapter 11 *Dodecahedron*,

$$\frac{5\Phi^5}{2\left(\Phi^2+1\right)}\,ds^3.$$

Converting this to the side of the r.t. we get:

$$Volume_{\text{dodecahedron}} = \frac{5\Phi^5}{2\left(\Phi^2+1\right)} \left(\frac{2}{\sqrt{\Phi^2+1}}\right)^3 rts^3$$

$$= \frac{5\Phi^5}{2\left(\Phi^2+1\right)} \left(\frac{8}{\left(\Phi^2+1\right)^{3/2}}\right) rts^3$$

$$= \frac{20\Phi^5}{\left(\Phi^2+1\right)^{5/2}} rts^3 = 8.908130915\, rts^3. \qquad (16.11)$$

The volume of each of the 12 "extra" pyramids is $\frac{1}{3}$ (area of pentagon) $\left(\overline{CV}\right)$ (height of each raised vertex off of the face of the dodecahedron), which is

$$\frac{1}{3}\left(\frac{5\Phi^2}{4\sqrt{\Phi^2+1}}ds^2\right)\left(\frac{\Phi}{\Phi^2+1}rts\right).$$

We only have one problem: one of our values is in terms of the dodecahedron side. We need to convert that to the side of the r.t.

$$\frac{5\Phi^2}{4\sqrt{\Phi^2+1}}ds^2 = \frac{5\Phi^2}{4\sqrt{\Phi^2+1}}\left(\frac{2}{\sqrt{\Phi^2+1}}\right)^2 rts^2$$

$$= \frac{5\Phi^2}{4\sqrt{\Phi^2+1}}\left(\frac{4}{\Phi^2+1}rts^2\right) = \frac{5\Phi^2}{\left(\Phi^2+1\right)^{3/2}}rts^2.$$

$$Volume_{1\,\text{'extra' pyramid}} = \frac{1}{3}\left(\frac{5\Phi^2}{\left(\Phi^2+1\right)^{3/2}}rts^2\right)\left(\frac{\Phi}{\Phi^2+1}rts\right) = \frac{5\Phi^3}{3\left(\Phi^2+1\right)^{5/2}}rts^3.$$

$$Volume_{12\,\text{'extra' pyramids}} = \frac{20\Phi^3}{\left(\Phi^2+1\right)^{5/2}}rts^3.$$

$$Volume_{\text{r.t.}} = Volume_{\text{dodecahedron}} + Volume_{\text{'extra'}}$$

$$= \frac{20\Phi^5 + 20\Phi^3}{\left(\Phi^2+1\right)^{5/2}}rts^3 = 12.31073415...\,rts^3. \qquad (16.12)$$

16.2. Surface Area of the Rhombic Triacontahedron

$$Surface\,area = 30\,(\text{area of each face})$$

$$= 30\left(\frac{2\Phi}{\Phi^2+1}rts^2\right)$$

$$= \frac{60\Phi}{\Phi^2+1}rts^2 = 26.83281573\,rts^2. \qquad (16.13)$$

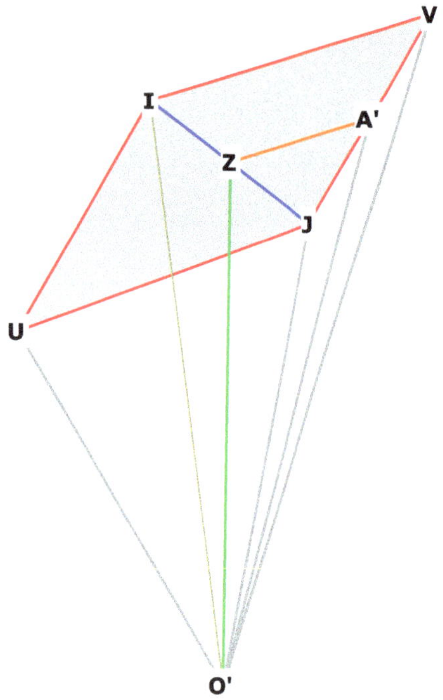

Figure 16.12: Showing the right triangle $\triangle O'ZA'$

16.3. Centroid Distances

Before we calculate the central and surface angles of the rhombic triacontahedron, let us complete our research into the distances from the centroid to various points of interest on this polyhedron.

We have already calculated the distances to the small and large axis vertices, and to the mid-face. Now we need to find the distance from the centroid to any mid-edge.

We are looking for $\overline{O'A'}$. We may write

$\overline{O'A'}^2 = \overline{O'Z}^2 + \overline{ZA'}^2$. We know $\overline{O'Z}$, but what is $\overline{ZA'}$.

$\overline{IZ} = $ one half $\overline{IJ}$ by construction.

$\overline{JA'} = $ one half $\overline{JV}$ by construction.

The triangles $\triangle VIJ$ and $\triangle ZA'J$ are congruent by angle-angle-angle. Triangle $\triangle VIJ$ is isosceles by construction, therefore triangle $\triangle ZA'J$ is isosceles and $\overline{ZA'} = \frac{1}{2}\,rts$.

$$\overline{O'A'}^2 = \left[\left(\frac{\Phi^2}{2}\right)\left(\frac{2}{\sqrt{\Phi^2+1}}\,rts\right)\right]^2 + \left(\frac{1}{2}rts\right)^2 = \left(\frac{\Phi^4}{\Phi^2+1} + \frac{1}{4}\right)rts^2$$

$$\overline{O'A'} = \sqrt{\frac{\Phi^4}{\Phi^2+1} + \frac{1}{4}}\,rts = 1.464386285\,rts. \tag{16.14}$$

Let us summarize:

Distance from centroid to long-axis r.t. face $(\overline{O'V}, \overline{O'U}, r_{outer}) = \Phi\,rts$ $\qquad$ (16.15)

Distance from centroid to short-axis r.t. face $(\overline{O'J}, \overline{O'I}, r_{inner}) = \dfrac{\sqrt{3}\Phi}{\sqrt{\Phi^2 + 1}} rts$ (16.16)

We know that $ds = \dfrac{2}{\sqrt{\Phi^2 + 1}}$, so

Distance from centroid to mid-r.t. face $(\overline{O'Z}) = \Phi^2 ds = \dfrac{\Phi^2}{2} \left(\dfrac{2}{\sqrt{\Phi^2 + 1}} rts \right)$

$$= \dfrac{\Phi^2}{\sqrt{\Phi^2 + 1}} rts.$$ (16.17)

Distance from centroid to mid-edge $(\overline{O'A'}) = \sqrt{\dfrac{\Phi^4}{\Phi^2 + 1} + \dfrac{1}{4}} rts.$ (16.18)

16.4. Central Angles of the Rhombic Triacontahedron

There are three central angles of the r.t. that are of interest. Refer to Figure 16.13.

The first is $\angle UO'V$, central angle of the long-axis. The second is $\angle IO'J$, central angle of the short-axis. The third and primary central angle is $\angle IO'V$, the central angle of each adjacent side.

$\overline{O'Z}$ is perpendicular to $\overline{IJ}$, and to the plane of the r.t. face, IUJV.

$\overline{O'Z}$ bisects $\overline{IJ}$ at Z.

Triangles $\triangle UO'Z$ and $\triangle VO'Z$ are right.

So we can write

$$\sin(\angle ZO'V) = \frac{\overline{ZV}}{\overline{O'V}} = \frac{\frac{\Phi}{\sqrt{\Phi^2+1}}}{\Phi}$$

$$rts = \frac{1}{\sqrt{\Phi^2 + 1}}$$

$$\angle ZO'V = 31.71747741°$$

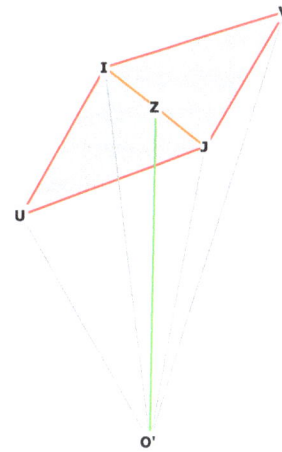

Figure 16.13: Rhombic triacontahedron internal pyramid

We recognize this angle (see Appendix B, *Phi Ratio Triangle*) as being part of a 1, Φ, $\sqrt{\Phi^2 + 1}$ ratio triangle.

The ratio's of these distances are shown below in Figure 16.14, which shows the relationship between $\overline{VZ}$, $\overline{O'Z}$, and $\overline{O'V}$:

V_____Z_____O'_____V
    1   Φ $\sqrt{\Phi^2+1}$

Figure 16.14

Now $\angle UO'V = 2(\angle ZO'V)$, so

$$\angle UO'V = 63.4349488°.$$ (16.19)

Let's find $\angle IO'J$, the central angle of the small axis of the r.t. face.

$\angle IO'Z$ and $\angle ZO'J$ are right. So we write

$$\tan(\angle IO'Z) = \frac{\overline{IZ}}{\overline{O'Z}}.$$

We know $\overline{IZ} = $ one-half the side of the dodecahedron, or $\frac{1}{2} ds$.

From above we know that $\overline{O'Z} = \frac{\Phi^2}{2} ds$. So

$$\tan(\angle IO'Z) = \frac{\frac{1}{2}}{\frac{\Phi^2}{2}} = \frac{1}{\Phi^2}. \angle IO'Z = 20.90515744°.$$

$$\text{Therefore} \angle IO'J = 2\left(\angle IO'Z\right) = 41.81031488°. \tag{16.20}$$

Note that $\frac{\overline{IZ}}{\overline{O'Z}} = \frac{1}{\Phi^2}$, so that $\overline{IZ}$ and $\overline{O'Z}$ have a relationship based on the square of Φ.

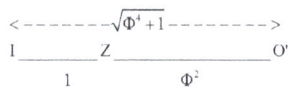

```
<-------√Φ⁴+1-------->
I_____Z_____O'
    1              Φ²
```

Figure 16.15: Showing the relationship between $\overline{O'Z}, \overline{IZ}$, and $\overline{O'I}$

This division is in ratio 1, Φ^2, $\sqrt{\Phi^4 + 1}$.

To find $\angle IO'V$, recognize that the point V lies in a straight line directly above the center of the dodecahedron face. So the angle $\angle IO'V$ is the same as the angle from I to O' to a point (G) in the middle of the dodecahedron face.

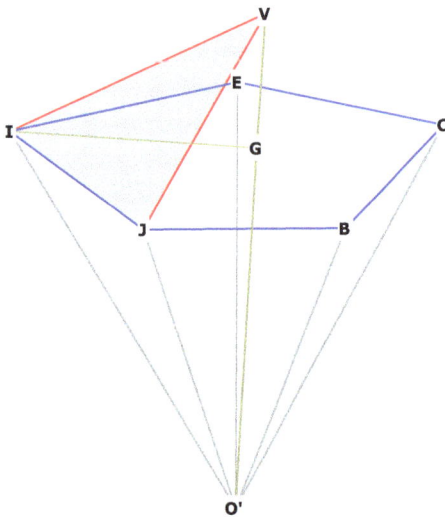

Figure 16.16: The central angle IO'V = angle IO'G

The triangle $\triangle O'GI$ is right by construction, so we need only to know $\overline{GI}$ and $\overline{O'G}$.

We know from *Area of the Pentagon* that $\overline{GI} = \frac{\Phi}{\sqrt{\Phi^2+1}}\, ds$, and from *Dodecahedron* we know the distance $\overline{O'G} = \frac{\Phi^3}{2\sqrt{\Phi^2+1}}\, ds$.

$$\text{So } \tan(\angle IO'G) = \frac{\overline{GI}}{\overline{O'G}} = \frac{\Phi}{\sqrt{\Phi^2+1}}\left(\frac{2\sqrt{\Phi^2+1}}{\Phi^3}\right) = \frac{2}{\Phi^2}.$$

$$\angle IO'G = \angle IO'V = 37.37736813^\circ. \tag{16.21}$$

16.5. Surface Angles of the Rhombic Triacontahedron

Observe from Figure 16.13 that the angles we are looking for are $\angle IVJ$ and $\angle UIV$.

Z is a bisector of $\overline{IZ}$ so the triangle $\triangle IVZ$ is right.

$\overline{IV} =$ the side of the r.t. $= rts$.

$\overline{ZV} = \frac{\Phi}{\sqrt{\Phi^2+1}}\, rts$, which we already found above.

So we can write

$$\cos(\angle IVZ) = \frac{\overline{ZV}}{\overline{IV}} = \frac{\Phi}{\sqrt{\Phi^2+1}}.$$

$$\angle IVZ = 31.71747441^\circ.$$

We recognize immediately, from *The Phi Right Triangle*, that this angle indicates a right triangle whose long and short sides are divided in Extreme and Mean Ratio (Phi ratio).

Therefore $\frac{\overline{ZV}}{\overline{IZ}} = \Phi$.

And so the ratio between the long axis $\overline{VU}$ and the short axis $\overline{IJ}$ (see Figure 16.13) must also be in Mean and Extreme Ratio.

Therefore $\frac{\overline{VU}}{\overline{IJ}} = \Phi$.

So the face of the rhombic triacontahedron is a Φ rhombus. The face angle we want, $\triangle IVJ$, is then twice $\angle IVZ$.

$$\angle IVJ = 63.43494882^\circ = \text{short axis face angle.} \tag{16.22}$$

One-half of the other face angle, $\angle ZIV$, is just $90^\circ - \angle IVZ = 58.28252558^\circ$.

So

$$\angle UIV = 116.5650512^\circ = \text{long axis face angle.} \tag{16.23}$$

16.6. Lengths of the long axis $\overline{VU}$ and the short axis $\overline{IJ}$ of the r.t. face

Refer to Figure 16.13

$$\overline{VU} = 2\left(\overline{ZV}\right) = \frac{2\Phi}{\sqrt{\Phi^2+1}}\, rts = 1.701301617\, rts. \tag{16.24}$$

$$\overline{IJ} = \text{side dodecahedron} = ds = \frac{2}{\sqrt{\Phi^2+1}}\, rts = 1.051462224\, rts. \tag{16.25}$$

As stated above,

$$\frac{\overline{VU}}{\overline{IJ}} = \frac{\frac{2\Phi}{\sqrt{\Phi^2+1}}\, rts}{\frac{2}{\sqrt{\Phi^2+1}}\, rts} = \frac{2\Phi}{2} = \Phi.$$

16.7. Dihedral Angle of the Rhombic Triacontahedron

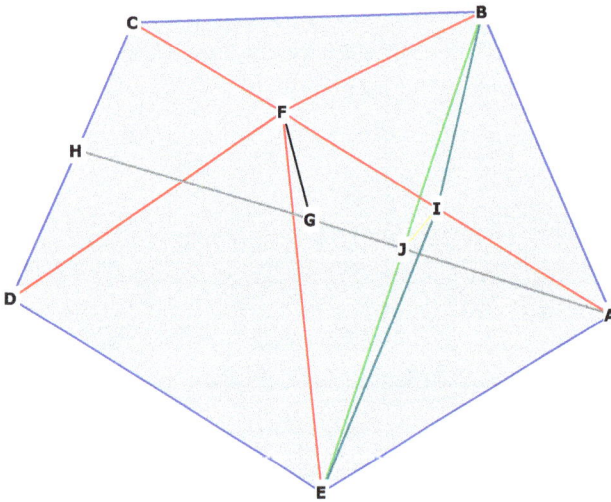

Figure 16.17: rhombic triacontahedron "cap"

We are looking for the dihedral angle, $\angle BIE$. Figure 16.17 shows an r.t. "cap" (in red) over the face of the dodecahedron (in blue), and shows that J and G are on $\overline{AH}$, the height of the pentagon. F is directly above the mid-face of the pentagon, at G. I is directly above J on $\overline{BE}$. $\angle BJI$ is right, allowing us to calculate $\angle BIJ$ or $\angle EIJ$. $\angle BIE$, the dihedral angle, is then just twice that.

$\overline{AH}$ and $\overline{BE}$ lie on the pentagonal plane ABCDE.

$\overline{BI}$ is a line along an r.t. face, $\overline{EI}$ is a line along another of the faces.

Here is our plan of attack:

We already know that $\angle BFA$ is $63.43494882°$. This can be rewritten without loss of information as $2\left[\sin^{-1}\left(\frac{1}{\sqrt{\Phi^2+1}}\right)\right]$. Triangle $\triangle BIF$ is right by construction. $\overline{BF}$ is the side of the r.t., or rts. Here we must use trigonometry to get $\overline{BI}$, by taking the sine of $\angle BFA$ so we can find $\overline{BI}$. $\overline{BE}$ is bisected by $\overline{AH}$, the angle bisector of $\angle BAE$ and a diagonal of the pentagon. $\overline{AJ}$ is known from *Pentagon Construction*. To get $\angle BIJ$, we can take the sine of $\angle BIJ = \frac{BJ}{BI}$. The dihedral angle $\angle BIE$ is twice angle $\angle BIJ$.

First let's find $\overline{BJ}$. $\overline{BJ}$ is one-half the diagonal of the pentagon with sides equal to the side of the dodecahedron. We must change units in terms of the side of the r.t. Remember that $ds = \frac{2}{\sqrt{\Phi^2+1}} \, rts$.

So we write

$$\overline{BJ} = \left(\frac{1}{2}\right)(\Phi)\left(\frac{2}{\sqrt{\Phi^2+1}} \, rts\right) = \frac{\Phi}{\sqrt{\Phi^2+1}} \, rts.$$

$$\sin\left\{2\left[\sin^{-1}\left(\frac{1}{\sqrt{\Phi^2+1}}\right)\right]\right\} = \frac{\overline{BI}}{\overline{BF}}.$$

$$\overline{BI} = (\overline{BF})\left(\sin\left\{2\left(\sin^{-1}\left[\frac{1}{\sqrt{\Phi^2+1}}\right]\right)\right\}\right).$$

$$\overline{BI} = rts\left(\sin\left[2*\sin^{-1}\left(\frac{1}{\sqrt{\Phi^2+1}}\right)\right]\right) = \frac{2\Phi}{\Phi^2+1} \, rts = 0.894427191 \, rts.$$

Now we can calculate the dihedral angle $\angle BIE$.

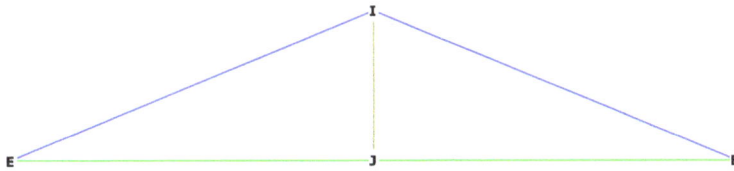

Figure 16.18: The dihedral angle BIE of the rhombic triacontahedron

We can write

$$\sin\left(\angle BIJ\right) = \frac{\overline{BJ}}{\overline{BI}} = \frac{\frac{\Phi}{\sqrt{\Phi^2+1}}}{\frac{2\Phi}{\Phi^2+1}} = \frac{\sqrt{\Phi^2+1}}{2}$$

$$\angle BIJ = 72°.$$

The dihedral angle $\angle BIE$ is $2\left(\angle BIJ\right)$.

$$\text{Dihedral Angle} = 144°. \tag{16.26}$$

16.8. Planar Distances

The distance from the centroid to any of the 12 long-axis vertices above the dodecahedron faces is $\Phi \, (rts.)$

The distance from the centroid to any of the 20 short-axis vertices is $\frac{\sqrt{3}\Phi}{\sqrt{\Phi^2+1}} \, rts = 1.473370419 \, rts.$

Consider and look at the diameter of the sphere which encloses the 12 long-axis vertices of the icosahedron. This line is shown in Figure 16.19 as $\overline{UO'W}$. The diameter passes through the centroid at O' and also through the middle of the two large pentagonal planes marked in Figure 16.19.

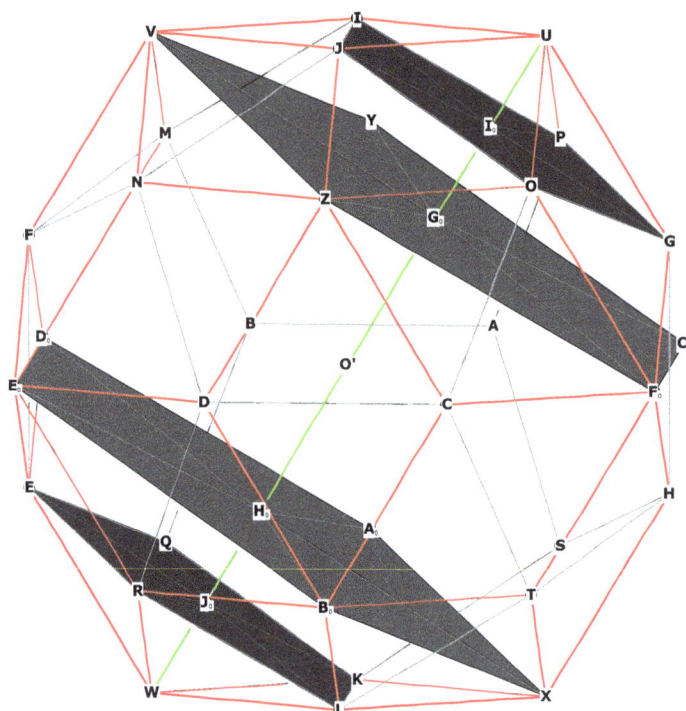

Figure 16.19: The internal pentagonal planes of the r.t.. (Figure 16.3) repeated

It also passes through the top and bottom faces of the dodecahedron. These faces are also marked in Figure 16.19.

There are 4 highlighted pentagonal planes along the diameter UW, as well as the midpoint O'.

We have the extra distance off the top plane of the dodecahedron to the vertex U, marked as $\overline{I_0U}$, and the extra distance off the bottom plane of the dodecahedron to the vertex W, marked as $\overline{J_0W}$.

What are the relationships between U, I_0, G_0, O', H_0, J_0, and W?

We have already found the distance off the plane of the dodecahedron face ($\overline{I_0U}$, $\overline{J_0W}$) to be $\overline{CV}$. $\overline{CV}$ we found to be $\frac{\Phi}{\Phi^2+1}$ rts. So $\overline{I_0U} = \overline{J_0W} = \frac{\Phi}{\Phi^2+1}$ rts.

Refer back to Figure 16.19.

Let's find the distance $\overline{UG_0}$, or the distance from U to the first large pentagonal plane VYC_0F_0Z.

This will be easy, because we know that the sides of this pentagonal plane are just the long axes of the r.t. We can form a right triangle from U to any one of the vertices of VYC_0F_0Z, to the center G_0. Let's take the right triangle $\triangle UG_0Z$. From *Construction of the Pentagon* we can get $\overline{G_0Z}$, it is just the distance from pentagon center to one of the vertices. Inspection of Figure 16.19 shows that $\overline{UZ}$ is just the long axis of the r.t. face, UJZO.

G_0 is the center of the large internal pentagon. U is directly above G_0.

The sides of the large pentagon (in green) are all long axes of the r.t. faces, indicated in red.

Triangle $\triangle UG_0Z$ is right by construction.

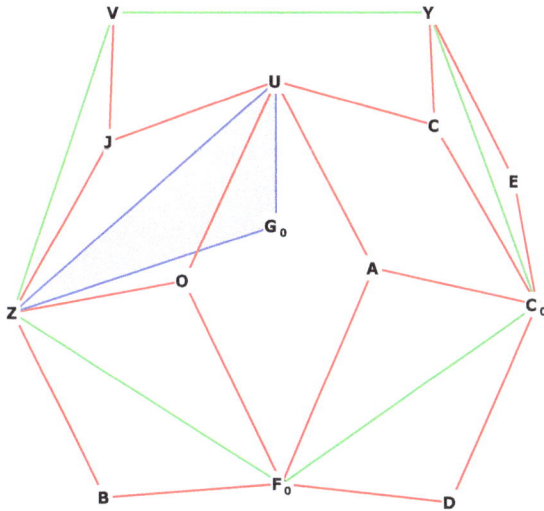

Figure 16.20: Showing the large internal pentagon VYC₀F₀Z and the right triangle UGₒZ.

$\overline{G_0 Z} = \frac{\Phi}{\sqrt{\Phi^2+1}}$ (side of pentagon), or long-axis of r.t. face.

$$\overline{G_0 Z} = \frac{\Phi}{\sqrt{\Phi^2+1}} \left(\frac{2\Phi}{\sqrt{\Phi^2+1}} \, rts \right) = \frac{2\Phi^2}{\Phi^2+1} \, rts.$$

$\overline{UZ} = $ side of pentagon, or long-axis of r.t. face $= \dfrac{2\Phi}{\sqrt{\Phi^2+1}} \, rts = 1.701301617 \, rts.$

Now we can find $\overline{UG_0}$:

$$\overline{UG_0}^2 = \overline{UZ}^2 - \overline{G_0 Z}^2 = \frac{4\Phi^2}{\Phi^2+1} - \frac{4\Phi^4}{(\Phi^2+1)^2}$$

$$= \frac{4\Phi^2 \left(\Phi^2+1 \right) - 4\Phi^4}{\left(\Phi^2+1 \right)^2}$$

$$= \frac{4\Phi^2}{(\Phi^2+1)^2} = \frac{4\Phi^2}{5\Phi^2} = \frac{4}{5} \, rts.$$

$$\overline{UG_0} = \frac{2}{\sqrt{5}} rts = \frac{2\Phi}{\Phi^2+1} \, rts = 0.894427191 \, rts. \tag{16.27}$$

So $\overline{UG_0} = 2 \left(\overline{UI_0} \right)$ and the plane VYC₀F₀Z of the dodecahedron is twice the distance from U as is the large internal pentagonal plane IPGOJ.

That means $\overline{UI_0} = I_0 G_0 = \frac{1}{2} \overline{UG_0} = \frac{\Phi}{\Phi^2+1} rts = 0.447213596 \, rts.$

16.8.1. WHAT IS THE DISTANCE BETWEEN THE CENTROID O′ AND THE PLANE IPGOJ, OR $\overline{O'G_0}$?

See Figure 16.19. It is $\overline{O'U} - \overline{UG_0}$.

We know $\overline{O'U}$, it is the radius of the outer sphere, or $\Phi \, rts$.

Therefore,

$$\overline{O'G_0} = \Phi - \frac{2\Phi}{\Phi^2+1} = \frac{\Phi^3+\Phi-2\Phi}{\Phi^2+1} = \frac{\Phi^3-\Phi}{\Phi^2+1} = \frac{\Phi^2}{\Phi^2+1} \, rts = 0.723605798 \, rts.$$

Now we have enough information to make our distance chart of internal planar distances of the rhombic triacontahedron, just as we did for the icosahedron and the dodecahedron.

16.8.2. DISTANCES BETWEEN INTERNAL PLANES OF THE RHOMBIC TRIACONTAHEDRON.

From the table of relationships (see Table 16.1) we see that:

$\overline{O'I_0}$ is divided in Extreme and Mean Ratio at Go_0.

$\overline{O'J_0}$ is divided in Extreme and Mean Ratio at H_0.

$\overline{UH_0}$ is divided in Mean and Extreme Ratio squared at G_0.

$\overline{WG_0}$ is divided in Mean and Extreme Ratio squared at H_0.

16.9. Conclusions:

The rhombic triacontahedron is a combined icosahedron-dodecahedron dual, so it is not surprising to see so many relationships based on the division in Mean and Extreme Ratio.

The rhombic triacontahedron contains all of the properties of the icosahedron and all of the properties of the dodecahedron and it tells us the proper nesting order of the 5 Platonic Solids.

16.10. Rhombic Triacontahedron Reference Tables

Table 16.1: Distances between Internal Planes of the Rhombic Triacontahedron. Graph reads vertically, by column. Let $\overline{UI_0}$ $\left(\text{which is}\ \frac{\Phi}{\Phi^2+1}\right) = 1$.

U		U			U
1					
I_0		2		I_0	
1					$\Phi^2 + 1$
G_0		G_0		Φ^2	
Φ					
O'		2Φ		O'	O'
Φ					
H_0		H_0		Φ^2	
1					$\Phi^2 + 1$
J_0		2		J_0	
1					
W		W			W

Table 16.2: Volume and Surface Area

Volume in terms of s	Volume in Unit Sphere	Surface Area in terms of s	Surface Area in Unit Sphere
12.31073415 s³	2.906170111 r³	26.83281573 s²	10.24922350 r²

Table 16.3: Angles

Central Angle			Dihedral Angle	Surface Angles	
long axis	short axis	adjacent vertex		long axis	short axis
63.4349488°	41.81031488°	37.37736813°	144°	116.5650512°	63.43494884°

Table 16.4: Centroid Distances

Centroid To		Centroid To Mid-edge	Centroid To Mid-face
short axis vertex	long axis vertex		
1.473370419 s	1.618033989 (Φ)	1.464386285 s	1.376381021

Table 16.5: Side to Radius

side / radius	
outer sphere	inner sphere (dodecahedron)
0.618033989 (Φ) rts	0.678715947 rts

Chapter 17

The Nested Platonic Solids

The nested Platonic Solids can be elegantly represented in the Rhombic Triacontahedron, as shown in *Rhombic Triacontahedron.*

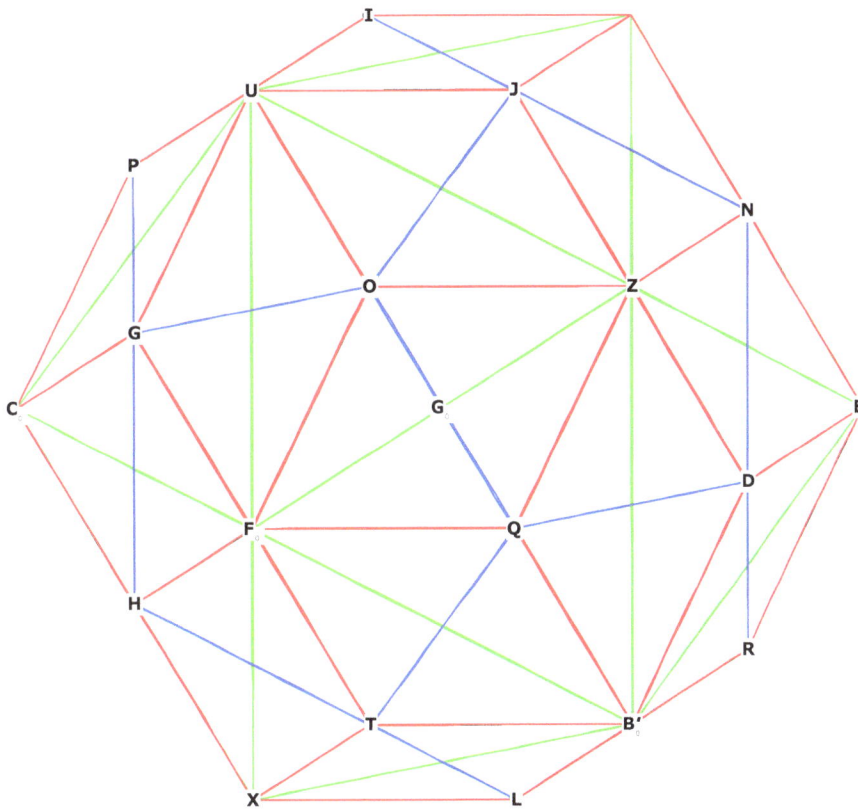

Figure 17.1: showing the rhombic triacontahedron (red) with its Phi ratio rhombi, icosahedron(green) with its equilateral triangle faces, and dodecahedron (blue) with its pentagonal faces

The Rhombic Triacontahedron is itself a combination of the Icosahedron and the Dodecahedron, and it demonstrates the proper relationship between the 5 nested Platonic Solids. You can see in Figure

17.1 how the edges of the dodecahedron are just the short diagonals of the rhombi of the rhombic triacontahedron, and the edges of the icosahedron are the long diagonals of each of the rhombi of the rhombic triacontahedron. Look at OF_0QZ, one of the rhombic faces of the rhombic triacontahedron. Long diagonal $\overline{ZF_0}$ in green is one of the edges of the triangular face F_0ZU of the icosahedron, and short diagonal $\overline{OQ}$ is one of the edges of the dodecahedron face OQTHG.

The vertices of the icosahedron are raised off the faces of the dodecahedron, so that the icosahedron surrounds the dodecahedron. If you look at point F_0 or Z, it rises slightly off the center of the dodecahedral face OQTHG or JNDQO.

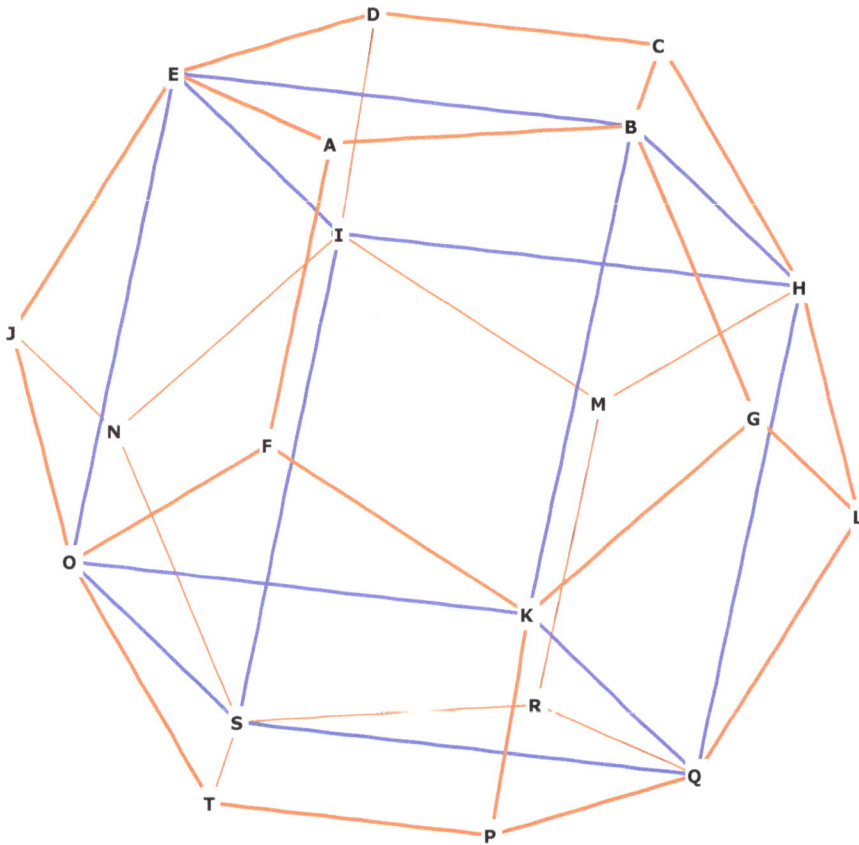

Figure 17.2: the cube fits directly on the vertices of a dodecahedron

The cube fits quite nicely within the dodecahedron, as shown in Figure 17.2. The cube has 8 vertices and 5 different cubes will fit within the dodecahedron. Each cube has 12 edges, and each edge will be a diagonal of one of the 12 pentagonal faces of the dodecahedron. Since there are only 5 diagonals to a pentagon, there can only be 5 different cubes, each of which will be angled 36 degrees from each other.

(Why is this? Because the diagonals of a pentagon are angled 36 degrees from each other)

The tetrahedron and the octahedron fit nicely within the cube, as shown above. Figure 17.3 shows that the octahedron is formed from the intersecting lines of the two interlocking tetrahedrons. The

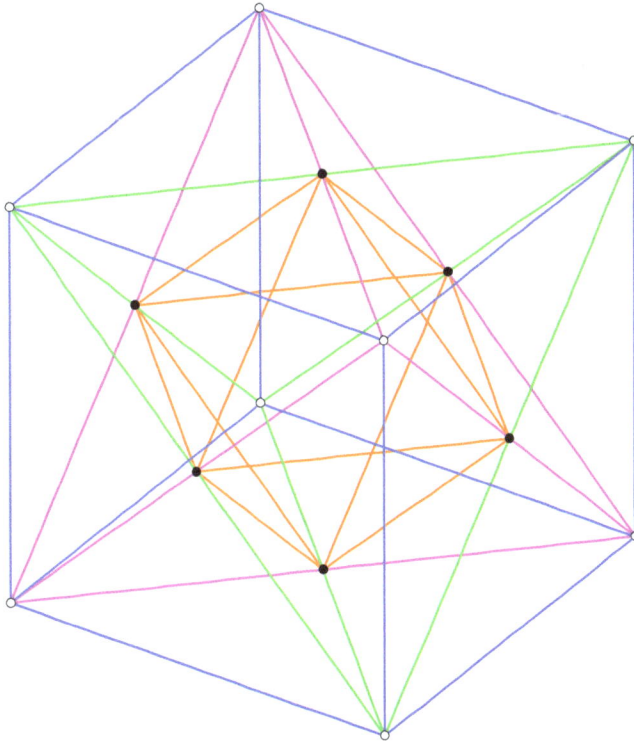

Figure 17.3: Star tetrahedron (two interlocking tetrahedra) fit inside the cube and define an octahedron

edges of the tetrahedrons are just the diagonals of the cube faces, and the intersection of the two tetrahedron edges meet precisely at the midpoint of the cube face. If you look at Figure 17.3 you'll see how the green and purple lines intersect precisely in the middle of the cube face, making an "X." At those points you will see one of the vertices of the octahedron.

The nesting order of the five Platonic Solids from largest to smallest, as given by the Rhombic Triacontahedron, is as follows:

- Icosahedron
- Dodecahedron
- Cube
- Tetrahedron
- Octahedron

Here's how the whole thing looks, all enclosed within a sphere (see Figure 17.4).

Surprisingly, even though there are 5 Platonic solids, there are only 3 different spheres which contain them. That is because the 4 vertices of each tetrahedron are 4 of the 8 cube vertices, and the 8 vertices of the cube are 8 of the 12 vertices of the dodecahedron.

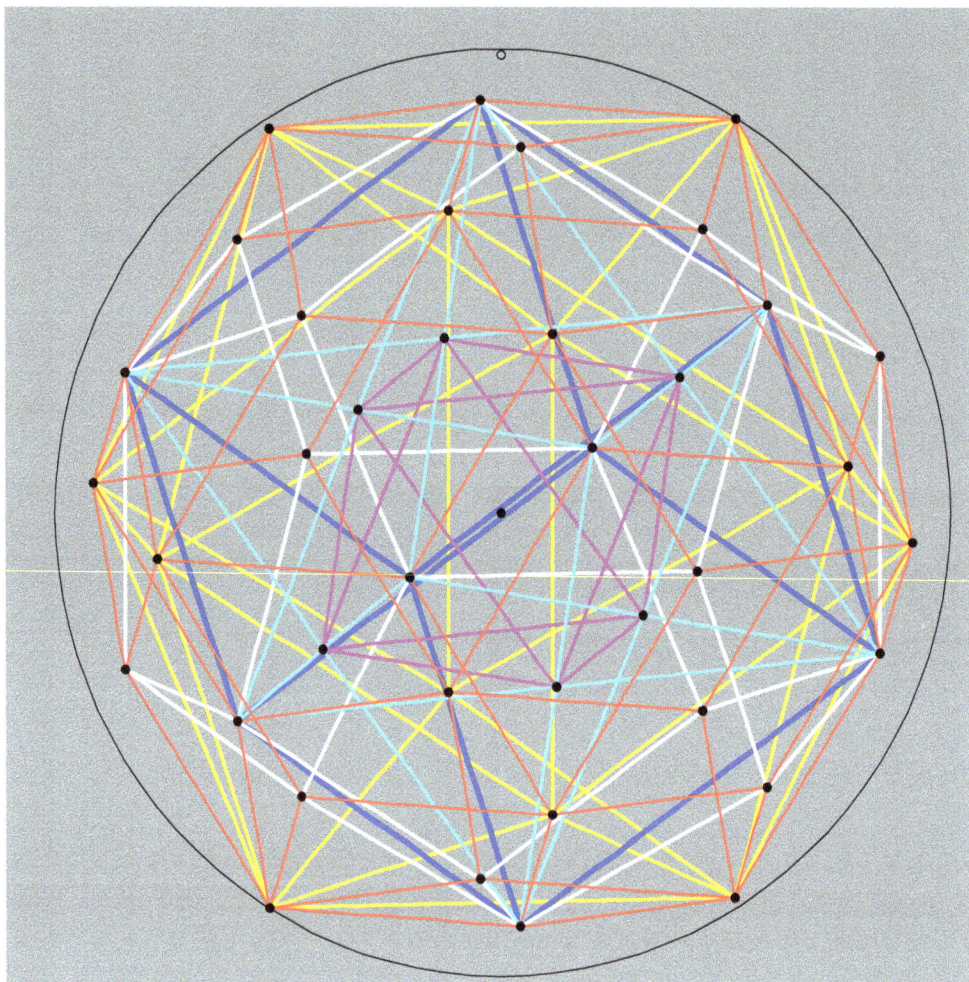

Figure 17.4: The 5 nested Platonic Solids inside a rhombic triacontahedron, surrounded by a sphere. The Icosahedron in yellow, the rhombic triacontahedron in red, the dodecahedron in white, the cube in blue, 2 interlocking tetrahedra in cyan, and the octahedron in magenta. Only the 12 vertices of the icosahedron touch the sphere boundary.

17.1. Radius of the Three Circumspheres

If we let the radius of the sphere that encloses the octahedron $= 1$, then what are the radii of the other two spheres?

Since the octahedron is formed from the midpoints of all of the cube faces, the sphere which encloses it fits precisely within the cube, as in Figure 17.5.

The radius of the circle that encloses the octahedron we will arbitrarily set $= 1$.

The next largest sphere encloses both the cube and the dodecahedron:

The radius of this sphere is $\sqrt{3}$ times the sphere that encloses the octahedron.

The first two spheres are the in-sphere of the cube and the circumsphere of the cube.

The outer sphere that encloses the icosahedron (see Figure 17.1) is slightly larger; $\frac{\sqrt{\Phi^2+1}}{\sqrt{3}}$ larger, in fact!

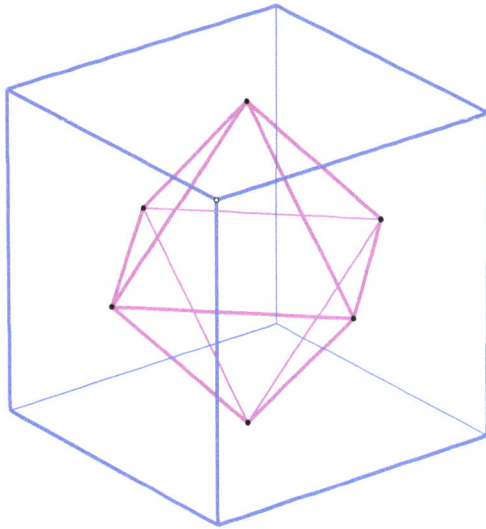

Figure 17.5: Octahedron in cube. Octahedron vertices are points in the middle of the cube faces

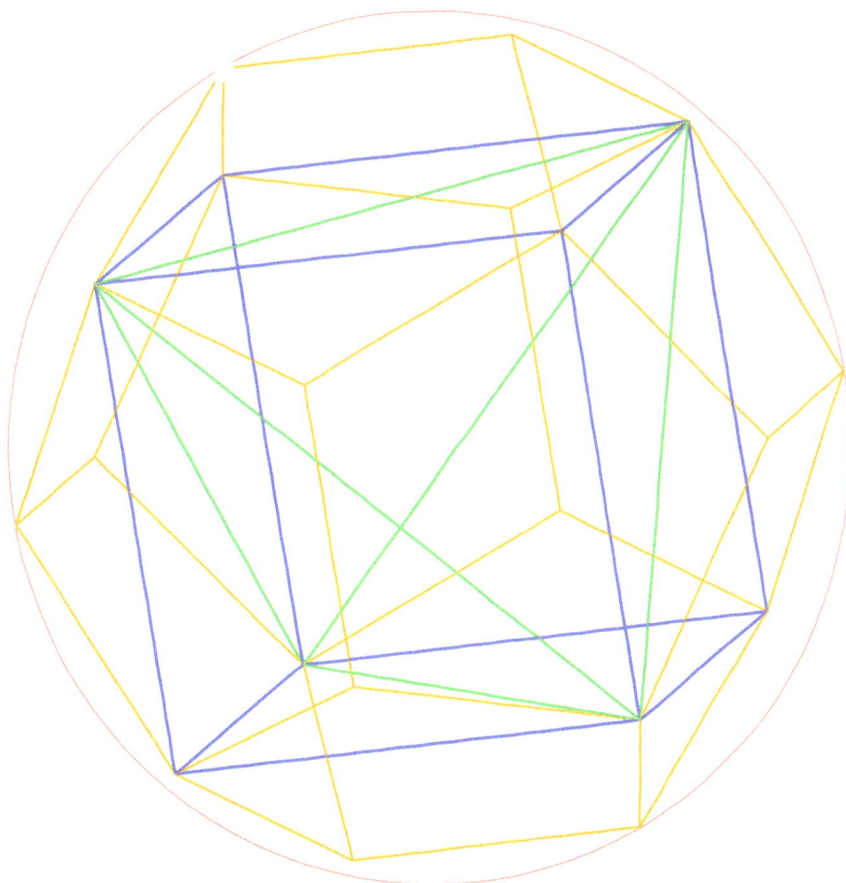

Figure 17.6: The sphere that encloses both dodecahedron and cube

So the radii of the three enclosing spheres is: $1, \sqrt{3}, \sqrt{\Phi^2 + 1}$.

Or, 1, 1.732050808, 1.902113033.

Here is a link to an animated GIF that shows all 5 Platonic Solids and the Rhombic Triacontahedron, rotating inside a sphere: http://www.kjmaclean.com/images/Nested.gif

Appendices

Appendix A — Combined Reference Charts

Volume and Surface Area

Polyhedron	Volume (s³)	Volume (r³)	Surface Area (s²)	Surface Area (r²)
Tetrahedron	0.11785113 s³	0.513200238 r³	1.732050808 s²	4.618802155 r²
Octahedron	0.471404521 s³	1.333333... r³	3.464101615 s²	6.92820323 r²
Cube	1.0 s³	1.539600718 r³	6.0 s²	8.0 r²
Icosahedron	2.181694991 s³	2.53615071 r³	8.660254038 s²	9.574541379 r²
Dodecahedron	7.663118963 s³	2.785163863 r³	20.64572881 s²	10.51462224 r²
Cube Octahedron	2.357022604 s³	2.357022604 r³	9.464101615 s²	9.464101615 r²
Rhombic Dodecahedron	8.485281375 r² (distance to 6 octahedral vertices)	3.079201436 s³	2.0 r³ (distance to 6 octahedral vertices)	11.3137085 s²
Icosa Dodecahedron	13.83552595 s³	3.266124627 r³	29.30598285 s²	11.19388937 r²
Rhombic Triacontahedron	12.31073415 s³	2.906170111 r³	26.83281573 s²	10.24922359 r²

Angles

Polyhedron	Central Angle (s)	Dihedral Angle	Surface Angle(s)
Tetrahedron	109.47122064°	70.52877936°	60°
Octahedron	90°	109.4712206°	60°
Cube	70.52877936°	90°	90°
Icosahedron	63.4349488°	138.1896852°	60°
Dodecahedron	41.81031488°	116.5650512°	108°
Cube Octahedron	70.52877936°	125.26438968° (between square and triangular faces)	90°
Rhombic Dodecahedron	Long axis: 90° Short axis: 70.52877936° Adjacent: 54.7356103°	120°	Long axis: 109.4712206° Short axis: 70.52877936°
Icosa Dodecahedron	36°	142.6226318°	Triangular faces: 60° Pentagonal faces: 108°
Rhombic Triacontahedron	Long axis: 63.4349488° Short axis: 41.81031488° Adjacent vertex: 37.37736813°	144°	Long axis: 116.5650512° Short axis: 63.4349488°

Centroid Distances

Polyhedron	Centroid to Vertex	Centroid to Mid-Edge	Centroid to Mid-Face
Tetrahedron	1.0 r	0.577350269 r	0.33333333 r
	0.612372436 s	0.353553391 s	0.204124145 s
Octahedron	1.0 r	0.707106781 r	0.577350269 r
	0.707106781 s	0.5 s	0.408248291 s
Cube	1.0 r	0.816496581 r	0.577350269 r
	0.866025404 s	0.707106781 s	0.5 s
Icosahedron	1.0 r	0.850650809 r	0.794654473 r
	0.951056517 s	0.809016995 s	0.755761314 s
Dodecahedron	1.0 r	0.934172359 r	0.794654473 r
	1.401258539 s	1.309016995 s	1.113516365 s
Cube Octahedron	1.0 r	0.866025404 r	To mid-triangle face: 0.816496581 r 0.816496581s
	1.0s	0.866025404 s	To mid-square face: 0.707106781 r 0.707106781 s
Rhombic Dodecahedron	1.0 r	0.829156198 r	0.707106781 r
	1.154700538 s	0.957427108 s	0.816496581 s
Icosa Dodecahedron	1.0 r	0.951056516 r	0.934172359 r
	1.618033989 s	1.538841769 s	1.511522629 s
Rhombic Triacontahedron	To short-axis vertex: 1.473370419 s	1.464386285 s	1.376381921 s
	To long-axis vertex: 1.618033989 s		

Side to Radius Ratios

Polyhedron	Side / radius
Tetrahedron	0.816496581
Octahedron	1.414213562
Cube	1.154700538
Icosahedron	1.051462224
Dodecahedron	0.713644179
Cube Octahedron	1.0
Rhombic Dodecahedron	To 6 cube vertices: 1.0 To 8 octahedral vertices: 1.154700538
Icosa Dodecahedron	0.618033989
Rhombic Triacontahedron	Outer sphere: 0.618033989 Inner sphere (dodecahedron) 0.678715947

Appendix B — The Equilateral Triangle

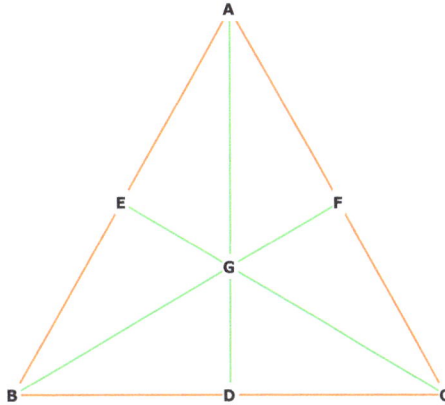

Given an equilateral triangle $\triangle ABC$ with side length = s,

Find $\overline{AG} = \overline{BG} = \overline{CG}$.

Find $\overline{EG} = \overline{DG} = \overline{FG}$.

Find the height of the triangle (h): $\overline{AD} = \overline{BF} = \overline{EC}$.

1. $\overline{AD}$ through G is an angle bisector of $\triangle BAC$, by construction. $\overline{AD}$ is a side bisector of $\overline{BC}$.

2. $\overline{CE}$ through G is an angle bisector of $\triangle BCA$, by construction. $\overline{CE}$ is a side bisector of $\overline{AB}$.

3. $\overline{BF}$ through G is an angle bisector of $\triangle ABC$, by construction. $\overline{BF}$ is a side bisector of $\overline{AC}$.

4. triangles $\triangle AEG, \triangle AFG, \triangle BEG, \triangle CEG, \triangle BDG, \triangle CDG$ are all congruent, by side-side-side and by construction.

Find h.

By the Pythagorean Theorem (PT),

$$h^2 = \overline{AC}^2 - \overline{DC}^2 = s^2 - \frac{1}{2}s^2 = s^2 - \frac{1}{4}s^2 = \frac{3}{4}s^2.$$

185

$$h = \frac{\sqrt{3}}{2} s. \tag{B.1}$$

Find the area of $\triangle ABC$.

The area of any triangle is $\frac{1}{2}$ (base) (height).

The base is $\overline{CA} = s$.

Area ABC $= \frac{1}{2} (s) \left(\frac{\sqrt{3}}{2} s \right) = \frac{\sqrt{3}}{4} s^2$.

Show each of the triangles in 4) above are similar to triangles $\triangle ADC$ and $\triangle ADB$.

5. $\overline{DC}$ is common to both triangles $\triangle DCG$ and $\triangle DCA$.

6. $\triangle ADC$ and $\triangle CDG$ are right, and are common to both triangles $\triangle DCG$ and $\triangle DCA$.

7. $\triangle DCG = \triangle DAC$ because both are bisections of equal angles $\angle BCA$ and $\angle BAC$.

8. Triangle $\triangle DCG$ is similar to triangle $\triangle DCA$ by angle-side-angle:

9. Therefore, $\overline{CG}$ is to $\overline{CD}$ as $\overline{AC}$ is to $\overline{AD}$, or,

$$\frac{\overline{CG}}{\overline{CD}} = \frac{\overline{AC}}{\overline{AD}} = \frac{\overline{CG}}{\frac{1}{2}} = \frac{1}{\frac{\sqrt{3}}{2}}.$$

$$2 \left(\overline{CG} \right) = \frac{2}{\sqrt{3}}.$$

$$\overline{CG} = \overline{AG} = \overline{BG} = \frac{1}{\sqrt{3}} s.$$

$$\overline{DG} = \overline{EG} = \overline{FG} = \frac{\sqrt{3}}{2} s - \frac{1}{\sqrt{3}} s = \frac{3-2}{\sqrt{3}} s = \frac{1}{2\sqrt{3}} s. \tag{B.2}$$

The longer portion of the diagonal is twice as long as the shorter part:

$$\frac{\overline{CG}}{\overline{DG}} = \frac{\frac{1}{\sqrt{3}}}{\frac{1}{2\sqrt{3}}} = 2. \tag{B.3}$$

B.1. Equilateral Triangle Reference Table

The Equilateral Triangle

Area	Height	Center to Vertex	Center to Mid-edge
0.433012702 s²	0.866025404 s	0.577350269 s	0.288675135 s

Appendix C — The Phi Right Triangle

Right triangles that have sides divided in Mean and Extreme Ratio have the following angles:

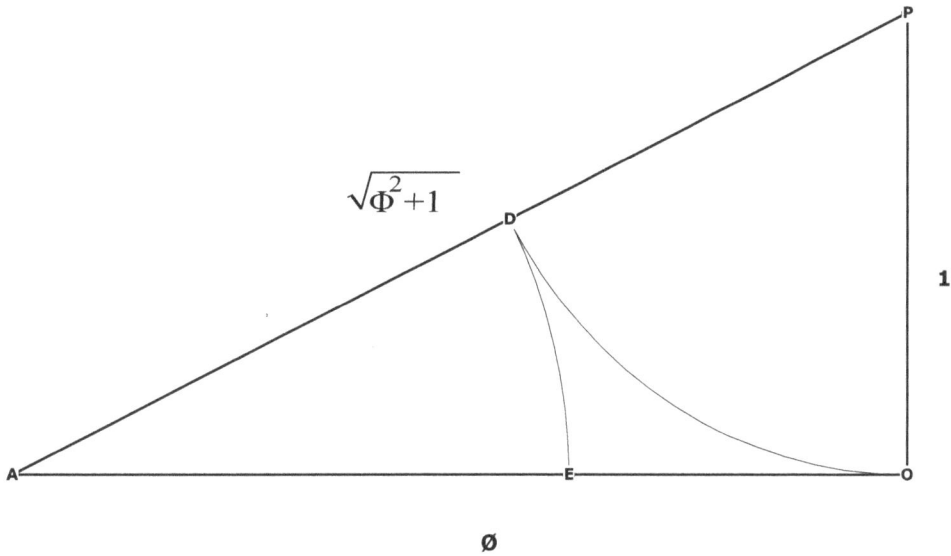

$$\tan(\angle OAP) = \frac{1}{\Phi}$$

$$\sin(\angle OAP) = \frac{1}{\sqrt{\Phi^2 + 1}}$$

$$\cos(\angle OAP) = \frac{\Phi}{\sqrt{\Phi^2 + 1}}$$

$$\angle OAP = 31.71747441° \quad \text{(C.1)}$$

$$\tan(\angle APO) = \Phi$$

$$\sin(\angle APO) = \frac{\Phi}{\sqrt{\Phi^2 + 1}}$$

$$\cos(\angle APO) = \frac{1}{\sqrt{\Phi^2 + 1}}$$

$$\angle APO = 58.28252559° \quad \text{(C.2)}$$

Appendix D — The Decagon

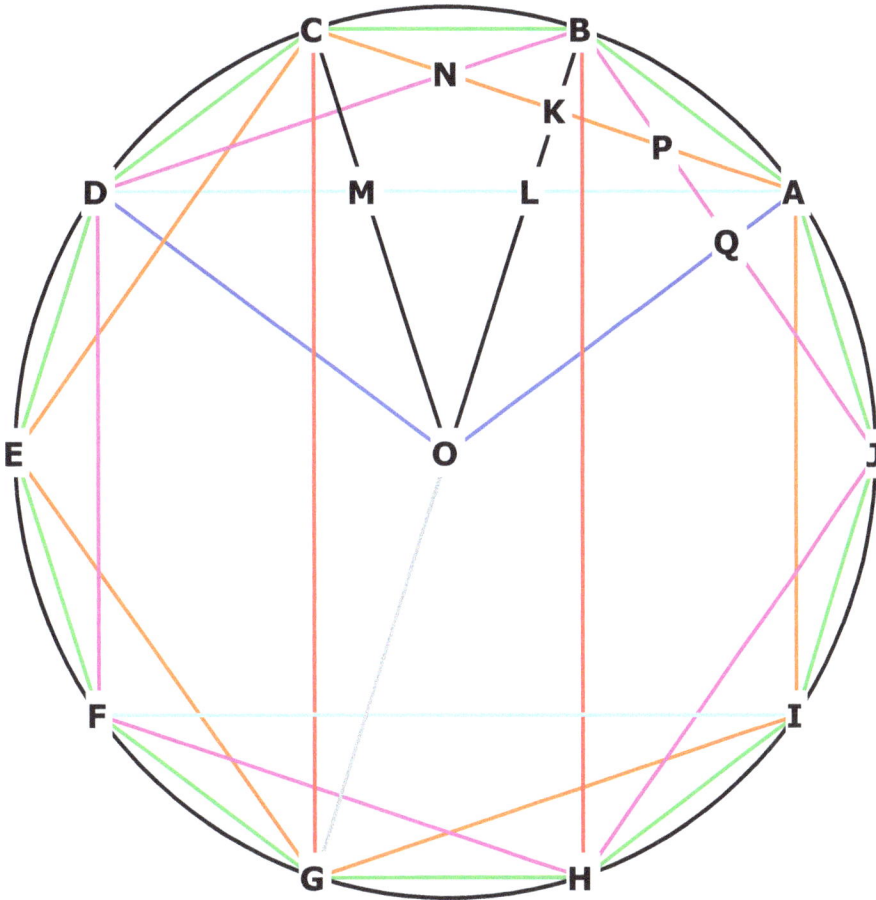

The decagon, showing two interocking pentagons

Notice that the decagon is composed of two interlocking pentagons, BDFHJ in purple, and ACEGI, in orange.

We already know the relationship between the radius of the pentagon and the pentagon side, so what is the relationship between the side of the decagon and the radius? Between the side of the decagon and the side of the pentagon?

D.1. Radius vs. Side of Decagon

We know that $\overline{OA} = \overline{OB} = \overline{OC} = \overline{OD}$ = radius of circle around decagon/pentagon.

We know, from *Construction of the Pentagon Part 2* that $\overline{OD} = \overline{OB}$ =radius, is the distance from center to a vertex of the pentagon, and that is $\left(\frac{\Phi}{\sqrt{\Phi^2+1}} \right) \cdot$ side of pentagon.

Inverting this, we know also that side of pentagon = $\left(\frac{\sqrt{\Phi^2+1}}{\Phi} \right) \cdot$ radius.

We also know from this that $\overline{OK} = \left(\frac{\Phi^2}{2\sqrt{\Phi^2+1}} \right) \cdot$ side of pentagon.

To find $\overline{AB}$, the side of the decagon, let's work with triangle $\triangle BKA$.

We know $\overline{AK}$ to be $\frac{1}{2} \left(\overline{AC} \right)$, because $\overline{OB}$ is a bisector of $\overline{AC}$.

$$
\begin{aligned}
\overline{BK} = \overline{OB} - \overline{OK} &= \frac{\Phi}{\sqrt{\Phi^2+1}} \cdot \text{pentagon side} - \frac{\Phi^2}{2\sqrt{\Phi^2+1}} \cdot \text{pentagon side} \\
&= \frac{2\Phi - \Phi^2}{2\sqrt{\Phi^2+1}} = \frac{2\Phi - \Phi - 1}{2\sqrt{\Phi^2+1}} \\
&= \frac{1}{2\Phi\sqrt{\Phi^2+1}} \cdot \text{side of pentagon.}
\end{aligned}
$$

(D.1)

Then

$$
\begin{aligned}
\overline{AB}^2 = \overline{AK}^2 + \overline{BK}^2 &= \frac{1}{4} + \frac{1}{4\Phi^2\left(\Phi^2+1\right)} \\
&= \frac{\Phi^2\left(\Phi^2+1\right)+1}{4\Phi^2\left(\Phi^2+1\right)} \\
&= \frac{\Phi^4 + \Phi^2 + 1}{4\Phi^2\left(\Phi^2+1\right)} = \frac{4\Phi^2}{4\Phi^2\left(\Phi^2+1\right)} \\
&= \frac{1}{\Phi^2+1}
\end{aligned}
$$

$$\overline{AB} - \text{side of decagon} = \frac{1}{\sqrt{\Phi^2+1}} \cdot \text{side of pentagon.} \tag{D.2}$$

$$\text{Side of pentagon} = \left(\sqrt{\Phi^2+1} \right) \cdot \text{side of decagon.} \tag{D.3}$$

So side of decagon is

$$\frac{1}{\sqrt{\Phi^2+1}} \left(\frac{\sqrt{\Phi^2+1}}{\Phi} \right) (\text{radius}) = \frac{1}{\Phi} (\text{radius.})$$

$$r = \Phi s, s = \frac{1}{\Phi} r. \tag{D.4}$$

Therefore the triangle $\triangle COB$, and all of the others like it, are Golden Mean Triangles.

D.2. $\overline{DA}$ in terms of the decagon side ds

(The pentagon side we will call ps.)

To get this, we work with triangle $\triangle DAI$ in Figure D.1. We know $\overline{DI}$, the diameter of the circle surrounding the decagon, or twice the radius. We also know $\overline{AI}$, the side of the pentagon.

So we can write:

$$\overline{DA}^2 = \overline{DI}^2 - \overline{AI}^2 = 4\Phi^2 - \left(\Phi^2 + 1\right) = 3\Phi^2 - 1 = \Phi^4.$$
$$\overline{DA} = \Phi^2 \, ds. \tag{D.5}$$

What is $\overline{CG} = \overline{BH}$, the diagonal of any of the inscribed pentagons? We will be working with triangle $\triangle BGH$.

We know $\overline{BG}$, the diameter or twice the radius, and we know $\overline{GH} = $ side of decagon.

$$\overline{CG}^2 = \overline{BG}^2 - \overline{GH}^2 = 4\Phi^2 - 1 = \Phi^4 + \Phi^2 = \Phi^2 \left(\Phi^2 + 1\right).$$
$$\overline{CG} = \left(\Phi\sqrt{\Phi^2 + 1}\right) ds. \tag{D.6}$$

D.3. Decagon Relationships

Let's start by showing that triangle $\triangle AOM$ is similar to triangle $\triangle DOL$.

- $\triangle DAF$ congruent to (the same as) $triangle ADI$ because they subtend equal arcs DF and AI.
- $\overline{OD} = \overline{OA}$ because they are both radii of the circle with center at O.
- $\overline{OM} = \overline{OL}$ because $\overline{ML}$ is parallel to $\overline{CB}$ and triangle $\triangle OML$ similar to triangle $\triangle OCB$ by angle-angle-angle.

  Therefore the triangles $\triangle AOM$ and $\triangle DOL$ are similar by angle-side-side.

Therefore $\overline{AM} = \overline{DL}$.

$\overline{AM} = \overline{DL}$ and both lines are divided by the equal distance $\overline{ML}$, therefore $\overline{DM} = \overline{AL}$.

Now we show that triangles $\triangle ALO$ and $\triangle DMO$ are congruent.

- $\overline{AL} = \overline{DM}$
- $\overline{DO} = \overline{AO}$ as they are both radii of the circle centered at O.
- $\overline{OM} = \overline{OL}$ as above.
- Therefore the triangles $\triangle ALO$ and $\triangle DMO$ are congruent by side-side-side.

Therefore $\overline{AL} = \overline{OL} = \overline{DM} = \overline{OM}$.

We can now write: $\overline{LM}$ is to $\overline{AL}$ as $\overline{AL}$ is to $\overline{AM}$.

Therefore the line $\overline{AM}$ is divided in Mean and Extreme Ratio at L.

And so the line $\overline{DL}$ is divided in Mean and Extreme Ratio at M, and $\overline{ML}$ also divides the line $\overline{DA}$ in Mean and Extreme Ratio at M and L. Now we can also write:

$\overline{AL}$ is to $\overline{AM}$ as $\overline{AM}$ is to $\overline{AD}$.

These are the same relationships and triangles we saw in the pentagon.

Is $\overline{AM} = \overline{AO}$**?**   Although we have a lot of information, we still haven't proven that $\overline{AM} = \overline{AO}$. We have shown that triangles $\triangle ALO$ and $\triangle DMO$ are congruent, but not that triangle $\triangle AMO$ is isosceles.

This is easily proven, because we know that $\overline{LM}$ is to $\overline{MO}$ as $\overline{CB}$ is to $\overline{CO}$, which is the division into Mean and Extreme Ratio.

But $\overline{MO} = \overline{AL}$, so $\overline{OM}$ is to $\overline{OC}$ as $\overline{AL}$ is to $\overline{AM}$, and therefore $\overline{AM} = \overline{OC}$.

So triangle $\triangle AOM$ is isosceles.

Because $\overline{OM}$ is to $\overline{AM}$ as $\overline{CB}$ is to $\overline{OC}$, $\overline{CB} = \overline{OM}$.

Therefore triangle $\triangle AOM$ is congruent to triangle $\triangle COB$.

Also we know that $\overline{CM} = \overline{BL} = \overline{LM}$.

We know that triangle $\triangle BAL$ is a golden mean triangle, because $\overline{BA} = \overline{AL}$, and $\overline{BL}$ is to $\overline{AB}$ as $\overline{CM}$ is to $\overline{CB}$, which is to say, a division in mean and extreme ratio.

$\angle CAD$ = one half that of $\angle BAD$, because it subtends an arc exactly one half as long.

Therefore $\overline{AC}$ is an angle bisector of $\angle BAD$, and $\overline{BK} = \overline{BL}$.

D.4. Calculating Decagon Relationships

Let's confirm all of these relationships with brute force calculation!

We will be working first with triangles $\triangle OKA$ and $\triangle AKL$.

$\overline{OB}$ is an angle bisector of $\angle COA$, so $\overline{OB}$ bisects $\overline{AC}$ at K, and $\overline{BK} = \overline{KL}$.

So $\overline{OB}$ is perpendicular to $\overline{AC}$.

We have previously found $\overline{BK}$ in relationship to the side of the pentagon. Let's convert this value to the side of the decagon:

$$\overline{BK} = \frac{1}{2\Phi\sqrt{\Phi^2 + 1}} \cdot \text{side of pentagon}$$
$$= \frac{1}{2\Phi\sqrt{\Phi^2 + 1}} \left(\sqrt{\Phi^2 + 1}\right) \cdot \text{side of decagon}$$
$$= \frac{1}{2\Phi}\, ds.$$

We also found previously that

$$\overline{OK} = \frac{\Phi^2}{2\sqrt{\Phi^2 + 1}} \cdot \text{side of pentagon}$$
$$= \frac{\Phi^2}{2\sqrt{\Phi^2 + 1}} \left(\sqrt{\Phi^2 + 1}\right) \cdot \text{side of decagon}$$
$$= \frac{\Phi^2}{2}\, ds.$$

$$\overline{AK} = \text{one half the side of the pentagon, or } \frac{1}{2}\left(\sqrt{\Phi^2+1}\right)\cdot\text{side of decagon}$$

$$= \frac{\sqrt{\Phi^2+1}}{2}\,ds.$$

To find $\overline{AL}$, write

$$\overline{AL}^2 = \overline{AK}^2 + \overline{KL}^2 = \frac{(\Phi^2+1)}{4} + \frac{1}{4\Phi^2}$$

$$= \frac{\Phi^2(\Phi^2+1)+1}{4\Phi^2}$$

$$= \frac{\Phi^4+\Phi^2+1}{4\Phi^2}$$

$$= \frac{4\Phi^2}{4\Phi^2} = ds. \tag{D.7}$$

$\overline{AL}$ = side of the decagon $\overline{BC}$, as we stated above.

What is $\overline{OL}$? If we are correct above, it should be equal to $\overline{AL}$ and $\overline{BC}$.

$\overline{OL} = \overline{OB} - \overline{BL}.$

$\overline{BL} = \overline{BK} + \overline{KL} = 2\left(\overline{BK}\right) = \frac{1}{\Phi}\,ds.$

$\overline{OL} = \Phi\,ds - \frac{1}{\Phi}\,ds = \left(\Phi - \frac{1}{\Phi}\right)ds = ds.$

Therefore $\overline{OL}$ = the side of the decagon = $\overline{AL} = \overline{BC}$.

The relationship of $\overline{OL}$ to $\overline{OB}$ is therefore

$$\frac{ds}{\frac{1}{\Phi}\,ds} = \Phi ds.$$

We can show $\overline{DM} = \overline{AL}$ with the same triangles on the upper left of the decagon.

Since we know $\overline{DA}$, we can write:

$$\overline{LM} = \overline{DA} - 2\left(\overline{AL}\right).$$

$$= \Phi^2\,ds - 2\,ds = (\Phi^2 - 2)\,ds = (\Phi + 1 - 2)\,ds = (\Phi - 1)\,ds$$

$$= \frac{1}{\Phi}\,ds. \tag{D.8}$$

So by calculation,

$$\overline{AM} = \overline{LM} + \overline{AL} = \frac{1}{\Phi}\,ds + ds = \left(1 + \frac{1}{\Phi}\right)ds = \Phi\,ds,$$

and $\overline{AM}$ is divided in Mean and Extreme Ratio at L.

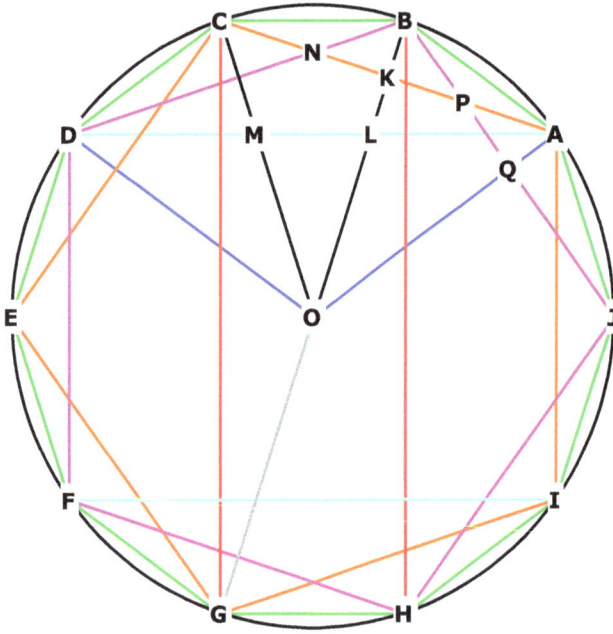

The relationship between $\overline{AM}$ and $\overline{AD}$. We said it was a Φ relationship. Let's see.

$$\frac{\overline{AD}}{\overline{AM}} = \frac{\Phi^2}{\Phi}\,ds = \Phi.$$

It isn't surprising that there are so many Phi relationships in the decagon, it being composed of two interlocking pentagons.

What about $\overline{AP}$, $\overline{BP}$, $\overline{KP}$, and $\overline{PQ}$? Are there any Phi relationships there?

First let's show that triangle $\triangle BKP$ is similar to triangle $\triangle BQO$.

- $\angle OBQ$ is common to both triangles.
- $\angle BKP$ and $\angle BQO$ are right.
- Therefore, by the property that all angles of a triangle must add to $180°$, $\angle BPK = \angle BOA$.

So the triangles are similar by angle-angle-angle.

Now we can determine $\overline{OQ}$ in triangle $\triangle BQO$ and from there determine $\overline{BP}$ and $\overline{KP}$. $\overline{BQ}$ is one-half the side of the pentagon.

$$\overline{OQ}^2 = \overline{OB}^2 - \overline{BQ}^2 = \Phi^2 - \frac{\Phi^2+1}{4} = \frac{4\Phi^2 - \Phi^2 - 1}{4} = \frac{3\Phi^2 - 1}{4} = \frac{\Phi^4}{4}$$

$$\overline{OQ} = \frac{\Phi^2}{2}. \tag{D.9}$$

Now we can write:

$\overline{BK}$ is to $\overline{BP}$ as $\overline{BQ}$ is to $\overline{OB}$, or, $\frac{\overline{BK}}{\overline{BP}} = \frac{\overline{BQ}}{\overline{OB}}$.

$$\overline{BP} = \overline{OB}\left(\frac{\overline{BK}}{\overline{BQ}}\right)$$

$$= \frac{\Phi\left(\frac{1}{2\Phi}\right)}{\frac{\sqrt{\Phi^2+1}}{2}} = \frac{1}{\sqrt{\Phi^2+1}}\,ds. \tag{D.10}$$

And, $\overline{OQ}$ is to $\overline{OB}$ as $\overline{KP}$ is to $\overline{BP}$, or, $\frac{OQ}{OB} = \frac{KP}{BP}$.

$$\overline{KP} = \overline{BP}\left(\frac{\overline{OQ}}{\overline{OB}}\right) = \frac{1}{\sqrt{\Phi^2 + 1}}\left(\frac{\frac{\Phi^2}{2}}{\Phi}\right)$$

$$= \frac{\Phi}{2\sqrt{\Phi^2 + 1}}\,ds. \qquad\qquad\text{(D.11)}$$

$$\overline{BN} = \overline{BP} \text{ and } \overline{KN} = \overline{KP}.$$

The relationship between $\overline{BP}$ and $\overline{KP}$ is $\frac{2}{\Phi}$.

It looks from the diagram that the sides of the pentagon, $\overline{BD}$, $\overline{BJ}$, and $\overline{AC}$, when they hit each other at N and P, divide the side of the pentagon equally in fourths. But this is not so.

Lets summarize our knowledge of the decagon in the following table.

Decagon Data

Identity	Value	Value (decimal representation)
$\overline{CB}$ = side of decagon	ds	1.0
$\overline{OB}$ = radius	$\Phi\,ds$	1.618033989
$\overline{AC}$ = side of pentagon	$\sqrt{\Phi^2+1}\,ds$	1.902113033
$\overline{AD}$ = long side of rectangle within decagon	$\Phi^2\,ds$	2.618033989
$\overline{CG}$ = diagonal of pentagon	$\Phi\sqrt{\Phi^2+1}\,ds$	3.077683538
$\overline{BG}$ = diameter	$2\Phi\,ds$	3.236067978
$\overline{AM}$	$\Phi\,ds$	1.618033989
$\overline{AL} = \overline{LO} = \overline{CB}$	ds	1.0
$\overline{LM} = \overline{BL}$	$\frac{1}{\Phi}ds$	0.618033989
$\overline{BK}$	$\frac{1}{2\Phi}ds$	0.309016994
$\overline{KL}$	$\frac{1}{2\Phi}ds$	0.309016994
$\overline{BL}$	$\frac{1}{\Phi}ds$	0.618033989
$\overline{BN} = \overline{BP}$	$\frac{1}{\sqrt{\Phi^2+1}}ds$	0.525731112
$\overline{KN} = \overline{KP}$	$\frac{\Phi}{2\sqrt{\Phi^2+1}}ds$	0.425325404

Appendix E — Phi Reduction

How does the mess

$$\frac{\sqrt{(16\Phi^2 - 3)} - \Phi\sqrt{(16 - 3\Phi^2)}}{8\Phi^2}$$

reduce elegantly to $\frac{1}{64}$?

This reduction shows the power of the Phi ratio to combine and divide into itself.

I could simply have used repeating decimals and avoided this complicated reduction. However, as I have remarked before, writing $\sqrt{(16\Phi^2 - 3)}$ is stating a PRECISE QUANTITY.

Writing 6.236067979... is sloppy. It is a repeating decimal and is only an approximation. It inherently results in loss of information. It is inelegant.

In mathematics it is always better to be precise than lazy.

When this reduction is performed on a calculator, round-off error and the imprecision of the repeating decimal yields accuracy to 7 decimal places. That is pretty good, but not perfect! Using repeating decimals yields $\frac{1}{63.99999994}$ instead of $\frac{1}{64}$.

Here is the reduction. First we square numerator and denominator:

$$\frac{\sqrt{(16\Phi^2 - 3)} - \Phi\sqrt{(16 - 3\Phi^2)}}{8\Phi^2} = \frac{(16\Phi^2 - 3) + \Phi^2(16 - 3\Phi^2) - 2\Phi\sqrt{(16\Phi^2 - 3)(16 - 3\Phi^2)}}{64\Phi^2}.$$

Now we will break down $\sqrt{(16\Phi^2 - 3)(16 - 3\Phi^2)}$.

$$\sqrt{(16\Phi^2 - 3)(16 - 3\Phi^2)} = \sqrt{256\Phi^2 - 48\Phi^4 - 48 + 9\Phi^2} = \sqrt{-48\Phi^4 + 265\Phi^2 - 48}.$$

Now note that $\Phi^4 = \Phi^3 + \Phi^2$ and $\Phi^3 = \Phi^2 + \Phi$ so that $\Phi^4 = 2\Phi^2 + \Phi$.

$$\sqrt{(16\Phi^2 - 3)(16 - 3\Phi^2)} = \sqrt{-96\Phi^2 - 48\Phi + 265\Phi^2 - 48} = \sqrt{169\Phi^2 - 48(\Phi + 1)}.$$

Now note that $\Phi^2 = \Phi + 1$ and

$$\sqrt{(16\Phi^2 - 3)(16 - 3\Phi^2)} = \sqrt{169\Phi^2 - 48\Phi^2}.$$

So

$$\sqrt{(16\Phi^2 - 3)(16 - 3\Phi^2)} = \sqrt{121\Phi^2} = 11\Phi.$$

Now we can write:

$$\frac{(16\Phi^2 - 3) + \Phi^2(16 - 3\Phi^2) - 2\Phi\sqrt{(16\Phi^2 - 3)(16 - 3\Phi^2)}}{64\Phi^2}$$

$$= (16\Phi^2 - 3) + \Phi^2(16 - 3\Phi^2) - 2\Phi(11\Phi)$$

$$= 16\Phi^2 - 3 + 16\Phi^2 - 3\Phi^4 - 22\Phi^2$$

$$= 32\Phi^2 - 3 - 3\Phi^4 - 22\Phi^2$$

$$= -3\Phi^4 + 10\Phi^2 - 3$$

$$= -3(2\Phi^2 + \Phi) + 10\Phi^2 - 3$$

$$= 4\Phi^2 - 3\Phi - 3$$

$$= 4\Phi^2 - 3(\Phi + 1)$$

$$= 4\Phi^2 - 3\Phi^2$$

$$= \Phi^2 !!!$$

Now we have to divide this by $64\Phi^2$, and we get

$$\frac{1}{64}.$$

At all times we have used exact mensuration. We have not approximated, we have not yielded one iota of error at any step in our calculations. That is the power of the division into mean and extreme ratio.

Bibliography

[1] Doczi, G. 1981. *The Power of Limits.* Boston: Shambhala.

[2] Densmore, D. & Heath, T.L. 2002. *Euclid's Elements.* Green Lion.

[3] Fuller, R. B. 1975. *Synergetics: Explorations in the Geometry of Thinking.* New York: Scribner.

[4] Ghyka, M. 1946. *The Geometry of Art and Life.* New York: Dover Books.

[5] Hart, G. W. The Encyclopedia of Polyhedra (Virtual Polyhedra).
http://www.georgehart.com/virtual-polyhedra/vp.html.

[6] Hart, G. W. & Picciotto, H. 2000. *Zome Geometry: Hands-on Learning with Zome^{TM} Models.* Zome Books.

[7] Herz-Fischer, R. 1998. *A Mathematical History of the Golden Number.* New York: Dover Books.

[8] Holden, A. 1991. *Shapes, Spaces, and Symmetry.* New York: Dover Books.

[9] Lawlor, R. 1989. *Sacred Geometry: Philosophy and Practice.* London: Thames & Hudson Ltd.

About the Author

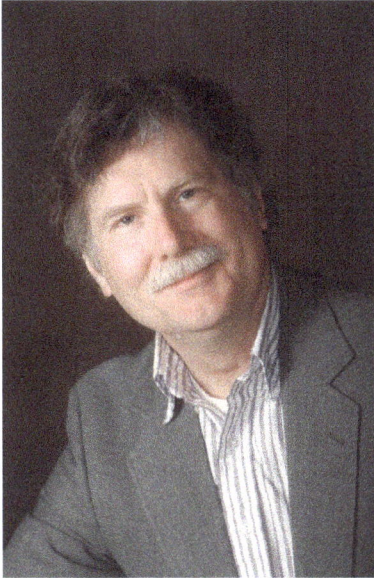

Kenneth MacLean is a writer and researcher living in Ann Arbor, Michigan, home of the Michigan Wolverines. He has authored seven books, dozens of essays, and four flash movies.

Contact: kmaclean@kjmaclean. com

Website: http://www.kjmaclean.com/

Other Books by the Author

The Vibrational Universe

The Manchild

The End of the Universe

Miracles Can Happen

Beyond the Beginning

Dialogues: Conversations with My Higher Self

Available in print or digital form at www.kjmaclean.com/Products/MainProductPage.php

www.ingramcontent.com/pod-product-compliance
Lightning Source LLC
Chambersburg PA
CBHW042107210326

41458CB00078B/6341